TECHNICAL REPORT

Adapting the Army's Training and Leader Development Programs for Future Challenges

James C. Crowley • Michael G. Shanley • Jeff Rothenberg • Jerry M. Sollinger

Prepared for the United States Army

T0306451

ARROYO CENTER

The research described in this report was sponsored by the United States Army under Contract No. W74V8H-06-C-0001.

Library of Congress Cataloging-in-Publication Data

Crowley, James C., 1945-
  Adapting the Army's training and leader development programs for future challenges / James C. Crowley, Michael G. Shanley, Jeff Rothenberg, Jerry M. Sollinger.
    pages cm
  Includes bibliographical references.
  ISBN 978-0-8330-7638-0 (pbk. : alk. paper)
  1. Command of troops—Study and teaching—United States. 2. Leadership—Study and teaching—United States. 3. United States. Army—Officers—Training of. 4. Military education—United States. I. Rand Corporation. II. Title. III. Title: Adapting the Army's ATLD programs for future challenges.

  UB413.C76 2013
  355.5'50973—dc23

                                                                              2012050181

The RAND Corporation is a nonprofit institution that helps improve policy and decisionmaking through research and analysis. RAND's publications do not necessarily reflect the opinions of its research clients and sponsors.

RAND® is a registered trademark.

Published 2013 by the RAND Corporation
1776 Main Street, P.O. Box 2138, Santa Monica, CA 90407-2138
1200 South Hayes Street, Arlington, VA 22202-5050
4570 Fifth Avenue, Suite 600, Pittsburgh, PA 15213-2665
RAND URL: http://www.rand.org/
To order RAND documents or to obtain additional information, contact
Distribution Services: Telephone: (310) 451-7002;
Fax: (310) 451-6915; Email: order@rand.org

Preface

The Army's operational requirements have expanded since the start of the 21st century. Its forces must be prepared to react to a wide range of potential missions, from peacekeeping to high-intensity conflict. At the same time, the Army must keep additional forces prepared while a significant proportion of its structure is deployed and operationally engaged. This new environment has created a need for major change in the Army's programs for training units and developing leaders. In 2010 RAND completed research designed to support Army efforts in these areas by identifying directions that the Army can follow to achieve the needed changes, and make those changes at a time when reduced budgets are likely. This report presents results of that research; it should interest those involved in designing Army training and leader development strategies and those involved in the process of providing resources for these strategies.

This research has been conducted in RAND Arroyo Center's Manpower and Training Program. RAND Arroyo Center, part of the RAND Corporation, is a federally funded research and development center sponsored by the United States Army. Questions and comments regarding this research are welcome and should be directed to the leaders of the research team, Jim Crowley or Michael Shanley, at crowley@rand.org and mikes@rand.org.

The Project Unique Identification Code (PUIC) for the project that produced this document is ATFCR09994.

For more information on RAND Arroyo Center, contact the Director of Operations (telephone 310-393-0411, extension 6419; FAX 310-451-6952; email Marcy_Agmon@rand.org), or visit Arroyo's Website at http://www.rand.org/ard/.

Contents

Figures

Tables

Summary

Background and Purpose

Trained units and competent leaders have always been, and remain, critical elements of Army operational success. The Army Training and Leader Development (ATLD) system is illustrated in Figure S.1. Six primary ATLD activities work in concert to have a direct role in achieving ATLD outputs, of trained units and competent leaders, with both of these outputs supporting the ultimate objective of near- and long-term operational readiness.

Figure S.1
Training and Leader Development Strategies, Primary Activities, and Outputs

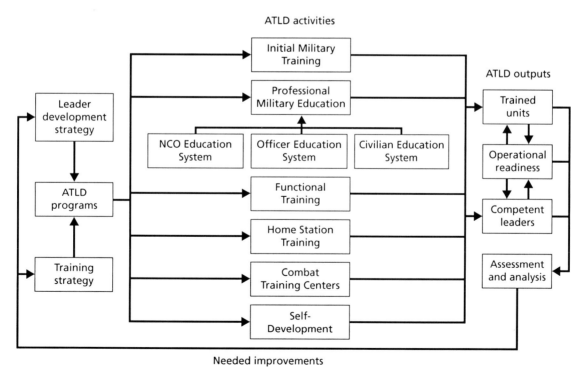

RAND *TR1236-S.1*

The six primary activities are the following:

- Initial Military Training (IMT) (top of Figure S.1) teaches soldiers and officers the tasks, supporting skills, and knowledge needed to be proficient at the first unit of assignment.
- Professional Military Education (PME) involves structured leader education courses.
- Functional Training involves courses designed to train specific functional tasks, skills, and knowledge.
- Home Station (HS) Training involves individual, leader, and collective training done at unit home stations.
- Combat Training Center (CTC) Training involves major collective training exercises conducted at a set of training centers with abundant resources.
- Self-development involves all of the learning activities done by the individual soldier for self-improvement.

The ATLD system is undergoing great change. The range of tasks and skills at which soldiers, leaders, and units must be proficient has increased. Traditionally, ATLD programs could focus training on the defeat of conventional enemy forces, but this is no longer the case. Now, Army units also must be able to defeat unconventional forces, develop partner forces, protect local populations, and support civil functions. At the same time, the requirement to have a significant portion of the operational force deployed is expected to be reduced but continue. Given these operational demands, the Army has implemented a process called Army Force Generation (ARFORGEN) to manage the preparation of its units through phases and provide regional commanders with the range of full-spectrum–capable forces needed to meet ongoing and contingency requirements.

While ATLD requirements have increased, future reductions in Army budgets will make it even more difficult to obtain funding for ATLD programs. This combination of increased requirements and limited resources means that ATLD strategies and programs must change as a part of a wider Army effort. To support needed changes, the Army has embarked on an Institutional Adaptation initiative, which contains three key elements:

- support of operational force readiness and ARFORGEN processes used to manage the force and ensure the ability to support demands for Army forces
- adopt an Enterprise Approach in which decisions are made for the overall good of the Army
- resource stewardship.

This initiative requires the institutional Army to re-examine its processes at a fundamental level and to make whatever changes are required to provide better support to the operational force.

The Department of the Army (DA)'s Director of Training and the U.S. Army Combined Arms Center's Deputy Commander-Training asked RAND's Arroyo Center to support these efforts. The study's primary objective is to identify directions that the Army can take to improve DA-level ATLD management processes and architectures.

While the major research for this report was completed in in 2010, follow-on research and coordination shows that the major findings, conclusions, and suggested directions with regard to adapting ATLD management processes remain valid and relevant.

Assumptions

We made three assumptions: First, that the Army will continue to deploy and conduct actual operations, but the level will decline to a point in which active units will have two or more years between deployments and reserve component units will have four to five or more. Second, that the Army will have to train units and develop leaders on a wide range of combat and non-combat skills and tasks. Finally, that there will be greatly increased pressure to reduce ATLD budgets.

Approach

We first examined ATLD processes in four key areas to understand how they function and to identify the organizations that participate in them. Based on the main Institutional Adaptation elements, we asked the following questions to assess the extent to which change to an Institutional Adaptation approach is warranted and is being achieved:

- **Support Operational Readiness and ARFORGEN Processes**. To what degree have management processes been adapted to support changing unit operational readiness requirements in the context of ARFORGEN processes?
- **Adopt an Enterprise Approach.** Do management processes focus on overall ATLD benefit and are they supported by structured assessment architecture?
- **Stewardship of Resources.** To what extent are decisions made after a systematic consideration of overall costs and benefits?

Based the answers to these questions, we developed conclusions about the adequacy of overall ATLD management processes, identified areas for improvement and developed directions that the Army could take to improve its ATLD management process.

Supporting Research Efforts

Our approach included the use of a new case study and three previous ATLD-related research efforts. As described above, a large number of different activities and a large number of Army organizations are involved in ATLD's management and execution. For this reason, an examination of a wide range of ATLD activities and management processes was necessary to develop valid conclusions about overall ATLD process improvement.

The case study examined a Training and Doctrine Command (TRADOC) course for junior leaders and focused on ATLD strategic management. The first previous research effort examined directions for improving the Army's Distributed Learning Program. The last two efforts focused on unit training and its support. Taken together, the four research efforts provide a reasonably broad basis for drawing conclusions about improving ATLD management processes.

Case Study: Advanced Leader Course (ALC). In the case study, we examined ATLD management in the context of a specific TRADOC Professional Military Education course: the Advanced Leader Course for junior noncommissioned officers (NCOs).

The Army Distributed Learning Program (TADLP). TADLP seeks to use multiple means and technologies to deliver training and learning to soldiers and leaders whenever and wherever it is needed. Distributed learning (DL) capabilities are increasingly important, because the time soldiers can spend in formal resident courses is becoming more limited. The purpose of this study was to assess TADLP performance in 2007 and 2008 and to outline options to improve its performance.

Brigade Combat Team Training Strategy Enablers. This study sought to help the Army identify options for improving the Army's training strategy for modernized Maneuver Brigade Combat Teams (BCTs).

Training Support for Operational Forces. The second unit training study, conducted between 2007 and 2008, focused on improving Training Support System (TSS) management processes. There are eleven specific TSS Management Decision Evaluation Package (MDEP) programs that support unit and institutional training. TSS management involves the programming, budgeting, and execution of TSS resources as a part of the Army's Planning, Programming, Budgeting, and Execution System (PPBES).

Conclusions

Based on our ALC case study and other ATLD-related research, we draw the following conclusions:

1. ATLD Programs Have Changed, But the Need for Major Change Remains

The Army is now entering an era in which it must be prepared to face a far wider range of possible missions and mission conditions than was the case in the 2001–2002 baseline period or, more currently, when the focus has been on counter-insurgency and stability operations in Iraq and Afghanistan. This situation widens and complicates training and leader development activities. As a result, training strategies and activities must change as well, and these changes must be made within the resources available to ATLD programs. ATLD programs have historically been funded at less-than-required levels, and there is no reasonable expectation that the level will increase. In fact, given the current Army budget outlook, the level is far more likely to decline.

2. Implementing Needed Changes Will Require Difficult Decisions

The changes needed are not a matter of going back to baseline strategies and programs. For example, full-spectrum scenarios at maneuver CTCs will be very different from the major conflict–focused scenarios of the baseline period or from the counter-insurgency–focused scenarios in recent CTC mission rehearsal exercises. The ATLD resources needed to support these new scenarios will change considerably.

Changes such as these will be needed across a wide range of training and leader development activities to meet a different balance of critical tasks, skills, and conditions. In a period of no-growth or declining budgets, increases in one area will inevitably require decreases in other. Many hard decisions will have to be made.

3. The Current ATLD Management Processes Are Not Set Up for Major Change, Nor Are They Flexible

Current ATLD management processes were developed to sustain and make incremental improvements to successful, well understood, and generally stable ATLD strategies. As a result, ATLD programs can be adapted in small increments. Future ATLD processes must also have increased near-term flexibility. The Army's efforts to adapt and meet emerging training requirements for operations in Iraq and Afghanistan were made possible because of major efforts, ad hoc processes, and the use of supplemental and operations funds, and a needed improvement is developing a system that is more capable of responding to new and unseen operational training support needs.

4. Better Integration of Training and Leader Development Strategies and Programs Is Needed

There are no systemic processes in place to integrate training and leader development strategies and programs for overall readiness benefit. At the strategic level, both a Training Strategy and a Leader Development Strategy have been developed by different Army organizations, DA and TRADOC, respectively.[1] Both have identified the desired aggressive ends (e.g., full-readiness, adaptable leaders) but are only beginning to come to grips with the difficult task of developing a consensus concerning feasible ways and means for reaching them. For each, a wide range of initiatives has been outlined but not how they fit together. Importantly, nor has the source of the time and resources for these initiatives been delineated.

Even within unit training strategies, there is a need for better integration. Gunnery strategies are developed by different organizations than are the broader Combined Arms Training Strategies, and overall those strategies outline a far more extensive set of activities than units generally are capable of executing.

Integrated, well-defined ATLD strategies are important inputs to effective ATLD program management. They outline what the individual programs are to achieve in the context of the overall readiness requirements and provide a basis for reasonable allocation of resources across activities and programs.

At the program level, current processes focus on individual ATLD programs with little consideration across the programs for overall benefit to readiness outcomes within available resources.

5. The Training Program Evaluation Group's (TTPEG's) MDEP System Makes It Difficult to Make Decisions in the Context of Overall ATLD Benefit

Unit proficiency and leader competencies are achieved through direct training and leader development activities, such as the ALC (and other PME) and CTC rotations. Thus the logical management focus would be on direct ATLD activities.

However, the Army's process for managing resources, using MDEPs, defines programs at a much finer level of granularity and in a way that makes it difficult to manage major shifts in resources to support changing ATLD priorities. A few MDEPs in the TTPEG are "direct" in the sense that they focus on a key direct training or leader development activity. Many more are "support" MDEPs in that they provide resources to many different activities, but the full range of resources allocated to direct ATLD activities is not directly visible. This makes it difficult to associate resources and costs at the activity level. Along with the large number of supporting MDEPs, this makes complex as well as time-consuming the process of coordinating, integrating, and justifying resources for direct ATLD activities.

---

[1]   This is based on a review of the *Army Training Strategy*, dated April 2011; *A Leader Development Strategy for a 21st Century Army*, dated November 2009; and TRADOC Pamphlet 525-8-3, *Army Training Concept, 2012–2020*, dated January 2011.

6. Lack of Data Hampers Effective Stewardship of ATLD Resources and the ATLD Community's Ability to Make a Case for Needed Resources

The lack of activity cost data, discussed above, is coupled with the general lack of activity benefit data-objective measures of the effects of an activity's effect on unit training readiness or leader competencies. The result is that ATLD management processes do not give managers the capability to make objective decisions about the effective allocation of ATLD resources. Instead, decisions are made from the perspective of individual programs and types of resources, and not the overall ATLD benefit.

Even more importantly, the Army has no system for objectively determining unit training readiness or leader competency levels, or ATLD areas in need of improvement. Thus the ATLD community, compared with the Manning and Equipping communities, lacks the ability to match resource levels with quantifiable readiness outcomes. The overall result is that the ATLD community lacks the data to make an objective case for the resources it needs.

7. Complexity and Lack of Integration Limit Operational Focus and Strategic Decisionmaking

The lack of a "big picture" view of ATLD program performance and needs, the focus on ATLD strategy components and individual programs, and complexity of the strategic management process all make it difficult to focus on overall readiness goals. Many decisions are made in terms of component strategies, programs, or MDEPs rather than in the context of what these mean to overall ATLD improvement.

These considerations also make it difficult to include effective operational force representation in the large number of advisory forums and councils that underpin the processes. Moreover, these considerations mean that the potential for effective collaboration between the institutional Army staffs and operational force commands for time-constrained budget and execution-year decisions is even more limited.

Areas for Improvement

Our overall conclusion is that current ATLD management processes, which were developed to manage incremental change, now require fundamental changes. Based on our research, we identified three interrelated, general areas for ATLD program management improvement. These areas align with the Institutional Adaptation goals:

- more direct understanding and focus on operational force needs
- increased integration across strategies, ATLD programs, and other Program Evaluation Groups (PEGs)
- development of a more structured cost-benefit approach to making ATLD program decisions.

What to Do?

Based on this research, we conclude that broader institutional adaptation could significantly improve ATLD management processes, and so we have developed a number of directions to move toward this goal. These directions represent conceptual approaches. The basic thrust is to improve analytical capabilities and strategic governance architectures.

1. Improve the Overall Analytical and Data-Collection Processes

An overall analysis process is needed to support effective ATLD programs adaptation to changing requirements and conditions. The six-step process we propose is shown in Figure S.2. The overall goal is for the ATLD community to have a common understanding and a synchronized plan for improving key elements of ATLD.

Step One: Document ATLD activity outputs and costs. The first step involves understanding and collecting data on current individual ATLD program costs. It also involves understanding the amount of training or learning that resulted from each activity. Improving ATLD programs requires an accurate understanding of existing major activities to establish a starting point (base case) from which changes can be initiated.

Step Two: Quantify unit and leader performance needs. This step involves a systematic data-collection and analysis effort to understand general and specific training and leader development areas, skills, and tasks needing improvement. To focus effort, changes in ATLD programs should be based on an informed understanding of areas of unit weakness.

Step Three: Identify and prioritize areas for unit training and leader development improvement. The third step (boxes in the middle of Figure S.2) involves a structured analysis process that identifies and prioritizes critical areas of needed ATLD improvement as objectively as possible. The understanding of current unit and leader performance needs gained from Step Two is an important input. But new and changed requirements, such as those generated by new equipment, organizations, concepts, or operational requirements must also be considered. In this regard effective collaboration with Combatant Commands (COCOMs)

Figure S.2
Proposed ATLD Analysis Process

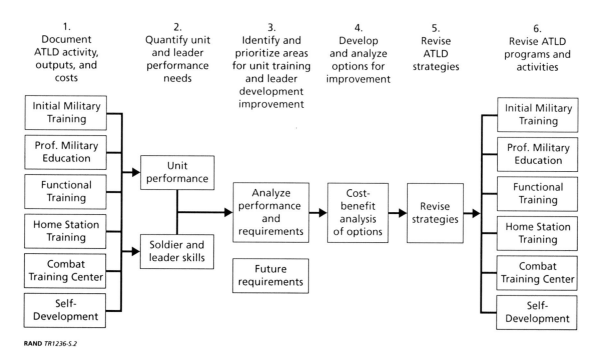

and Army Service Component Commands (ASCCs) to understand their changing operational needs and what these mean to prioritizing ATLD goals and objectives is highly important.

Step Four: Develop and analyze options for improvement. This requires a cost-benefit approach. It is perhaps the most complex step, because unit training readiness and leader competencies result from many activities, and effective improvement options often will require a multi-activity approach.

Step Five: Revise ATLD strategies. This step involves revising overall ATLD goals and objectives, not only in terms of what is of increased importance, but also what is less so. It also involves defining broadly, but with sufficient specificity to understand how ATLD activities should be reshaped and how resources should be reallocated.

Step Six: Revise ATLD programs and activities. This final step will require difficult decisions. In most cases, resources for enhancement in one area must be taken from other areas. Implementation plans should include a process for continued assessment and revision; the process we outline is iterative, not linear.

2. Improve Systems for Collecting Data on ATLD Achievements, Nature, and Needs

Making the best possible decisions on training and leader development programs and policies, program design, where to invest resources, and the level of investment needed requires a clear understanding of the nature of these programs. They also require data. Currently, data to inform such an understanding either do not exist or are not easily accessible. However, it is possible to set up a data-collection effort by taking advantage of ongoing efforts.

For the data that exist, the challenge is to establish systematic and coordinated collection. Many organizations are already collecting relevant data, but it is often difficult to get these into decisionmaking processes.

3. Create Improved Mechanisms for Managing by Direct ATLD Activity

As discussed earlier, the MDEP system complicates processes for managing by ATLD activity. Stronger mechanisms for cross-MDEP and cross-PEG management and visibility are needed, and DA could modify current program management mechanisms to enable management by primary training activity.

Management by direct ATLD activity would require modifying the current MDEP system to show the level of support provided by each activity from the range of supporting MDEPs. Under this system, MDEPs supporting direct training activities would be aligned directly under the activity they support, providing greater visibility to the degree of support for each. Note that supporting MDEPs would thus be affiliated with multiple direct training activities.

4. Unify Responsibility for Data Collection and Analysis and for Supporting ATLD Strategy and Program Management

Given the deficiencies in data and analysis capabilities that we have noted, and the complexity and difficulty of adapting the current training and leader development system, we believe a centralized effort will be required to fill in the gaps. We recommend forming a single, permanent staff organization to provide data and analysis support for ATLD strategy and program management.

While data collection and analysis are the two primary tasks required, they must be performed in an integrated, iterative process, as outlined in Figure S.2 above. Moreover, the data collected and functions performed should be brought into an overarching ATLD information technology (IT) system as quickly and as completely as possible.

The most feasible direction for improvement would be to establish within TRADOC, out of existing staff resources, a single organization for ATLD data collection and analysis. The majority of the organizations supporting the functions described above exist within TRADOC. TRADOC also owns a large portion of the Army's analytic capability, and has responsibility for integrating Doctrine, Organization, Training, Material, Leadership and Education, Personnel, and Facility (DOTMLPF) capabilities.

While the proposed organization would fall under TRADOC and support its current training and leader development roles, it would have a charter giving it a "direct support" relationship to the DA Director of Training to support its training and leader development policy and programming responsibilities.

5. Enhance ATLD and Army-Wide IT Architectures to Improve Data Collection and Analysis

An improved IT architecture would provide better support to ATLD analytical processes by increasing the amount of information available and by reducing the workload of collecting and analyzing that information. For a variety of reasons, we do not see any potential for a large near-term improvement, but taking steps to synchronize and focus the ongoing development of the Army Training Information System could result in improved support. To obtain broader support for the longer term, these efforts could move in the direction that IT architecture throughout the Department of Defense (DoD) is evolving toward a generic new approach known as service-oriented architecture (SOA), which holds the potential to provide greatly improved interoperability among systems.

6. Evolve Emerging ATLD Governance Structures and Processes to Improve Focus on Operational Force Readiness

Our research indicates a need to revise the strategic architecture to support ARFORGEN processes more effectively, to involve U.S. Army Forces Command (FORSCOM) and other unit-owning commands in ATLD decision processes, and to achieve a better balance across ATLD programs.

We outline three major directions for improving strategic governance:

- Re-establish an over-arching Training and Leader Development General Officer Steering Committee (GOSC) to support Army efforts to integrate training and leader development strategies and programs.
- Formalize FORSCOM's role as the Army's Readiness Core Enterprise to include a stated role of representing the operational force with a commensurate level of authority for influencing decisions and recommendations.
- Streamline governance forums to increase the ability of the operational force to effectively contribute to ATLD processes.

Implementing Suggested Changes

The directions we suggest would require major change, but are reasonable, feasible, and would provide significant improvement. We realize that reasonable alternatives exist, but the overall point is that significant changes in ATLD management processes are needed to focus more directly on operational force readiness and to foster resource stewardship through a more objective, analytical, cost-benefit approach. Compromise and rational risk assessment will be required to shape and resource revised ATLD strategies, and an objective cost-benefit approach to developing and selecting options for improvement will be key to making these decisions. Once in place, such an approach could better support both Program Objective Memorandum planning and short-term ATLD decisions, such as reacting to changed budget allocations in the budget and execution years.

- Moving in these directions could start in the near term and provide benefits, with an incremental implementation approach being used to continually improve data-collection and analysis processes. Moreover, the Army can and should make this incremental movement using existing organizational resources, without needing to add to them.

Broader Implications

Our examination reinforces the obvious conclusion that achieving needed training and leader development levels involves decisions and actions both inside and outside the TTPEG. Manning, equipping, and installation policy and programming decisions affect training and leader development, and resources from all PEGs provide critical support to ATLD activities. Goals among PEGs can conflict and require difficult decisions about what is best overall for the Army enterprise, so cross-PEG coordination is needed, with an especially important area being synchronization between the Training and Manning PEGs.

A reasonable argument is that the current operating environment has increased the scope of training and leader development requirements, justifying increased claims on resources. Nevertheless, the ATLD community has historically had difficulties presenting objective analysis to support balanced resource decisions among training, manning, and equipping functions. Absent such analysis, the result can be decisions to accept risks in training, because there is no real way to display analytically the effects on readiness. The training and leader development community must be able to make its case in a way that better informs the leadership of the risks and rewards of the hard decisions needed to take a synchronized Army Enterprise view across PEGs. This will, in turn, require that decisions across all PEGs consider training and leader development impacts and needs, and also that TTPEG decisions must consider broader readiness needs.

Acknowledgments

This project was sponsored by BG Richard Longo, the DA Director of Training, and BG Paul Funk, Deputy Commanding General for Training U.S. Army Combined Arms Center (CAC). Without their support and that from members of their staff, this study would not have been possible.

From Headquarters Department of the Army, BG (Ret.) Thomas Maffey, COL Lawrence Smith, Robert Parry, Harry Crumling, Brenda Granderson, Lee Jorde, Dean Camarella, Mary Ellen McCrillis, John Hughes, Ron Schexnayder, Charles Scott, Tom Macia, LTC Darran Anderson, Joe Kenny, Alan Young, and Jon Berlin in the Directorate of Training and COLs Nicholas Amodeo, John Keeter, and Timothy Burns, and Victoria Calhoun in the DA Office of Business Transformation provided valuable support.

The same is true for key contacts at U.S. Army Training Support Center (ATSC), including Terry Faber and Larry Matthews, who advised us and provided valuable support throughout the project.

Elsewhere within the Army, we received considerable input from Alan Craig, Gerald Purcell, and Ralph Steinway at the Headquarters of the Department of the Army (HQDA) G-1. They and other staff at G-1 and ASM Research, Inc. provided us with documentation and data from the Army Training Requirements and Resources System (ATRRS), as well as patiently answered our many questions about the data and the associated training processes they support. Special thanks are also due to Donald Chung and Ronald Weaver at TRADOC G-8, and David Doctor at ATSC, for the time they spent educating us about the Army's Individual Training Resource Model and, more generally, the Army's resourcing process.

At CAC, Denny Tighe, Clark Delavan, Matt Belford, Keith Beurskens, Rob Schwartzman, and Thomas Ryan provided thoughtful suggestions on a preliminary document. We are also indebted to Nate Godwin, Al Sutherland, Victor Macias, SMA Terry Sato, COL Kirk Palan, and Russ Hummel at U.S. Army Forces Command.

The Army's ALC for NCOs was a focal point of our study effort. Jeffrey Colimon, Danny Hubbard, and Jonathon "Dusty" Rhodes at Headquarters TRADOC, CSM Dean Keveles at the Fires Center of Excellence, CSM Raymond Chandler and SMA Richard Rosen at the Sergeants Major Academy, 1SG William Kerns at the Maneuver Center of Excellence, and 1SG Dean Francis at the Ordnance Center and School all helped us to gain the information and insights we needed in this key aspect of our work.

This work also benefited from the support of Frank Camm, Kristin Leuschner, Tom Lippiatt, Bryan Hallmark, and Chip Leonard. LTC Steven Cram spent considerable time supporting the project team by educating us concerning the Army's resourcing process while serving as a RAND Army Fellow.

Finally, we wish to thank our two reviewers for their careful, helpful comments on earlier drafts of this report. We benefited greatly from their feedback and from the assistance provided by all the individuals listed above. Errors of fact or interpretation, of course, remain the authors' responsibility.

Abbreviations

1SG	First Sergeant
ABO	Army Budget Office
AC	active component
ACOM	Army Command
ADP	Army Doctrinal Publication
AEB	Army Enterprise Board
ALC	Advanced Leader Course
ALDP	Army Leader Development Program
ALMS	Army Learning Management System
AMC	Army Materiel Command
ANCOC	Advanced NCO Course
APE	Army Program Element
AR	Army Regulation
ARCIC	Army Capabilities Integration Center
ARFORGEN	Army Force Generation
ARI	Army Research Institute
ARM-G	Automated Requirements Model–Guard
ARM-R	Automated Requirements Model–Resources
ARNG	Army National Guard
ARNGB	Army National Guard Bureau
ARPRINT	Army Program for Institutional Training
ASA	Assistant Secretary of the Army
ASAT	Automated Systems Approach to Training
ASCC	Army Service Component Command
ASTRM	Automated Strength Requirement Model
ATED	Army Training and Education Development
ATIS	Army Training Information System
ATLD	Army Training and Leader Development
ATLDC	Army Training and Leader Development Conference

ATRRS	Army Training Requirements and Resources System
ATSC	Army Training Support Center
AWS-M	analytical workspace model
BARS	BNCOC Automated Reservation System
BCKS	Battle Command Knowledge System
BCT	Brigade Combat Team
BCTC	Battle Command Training Center
BCTP	Battle Command Training Program
BNCOC	Basic Noncommissioned Officer Course
BRP	Budget Requirements Process
CAC	Combined Arms Center
CAD	Course Administrative Data
CAL	Center for Army Leadership
CALL	Center for Army Lessons Learned
CASC	Combat Arms Support Command
CATS	Combined Arms Training Strategies
CE	Core Enterprise
CES	Civilian Education System
CFX	command field exercises
CIO	Chief Information Officer
CoC	Council of Colonels
COCOM	combatant command
CSA	Chief of Staff of the Army
CSM	Command Sergeant Major
CTC	Combat Training Center
DA	Department of the Army
DAMO-FM	Army G-3/5/7 Force Management Directorate
DCS	Deputy Chief of Staff
DIMHRS	Defense Integrated Military Human Resources System
DL	distributed learning
DoD	Department of Defense
DOT	Director of Training
DOTMLPF	Doctrine, Organization, Training, Material, Leadership and Education, Personnel, and Facilities
DRB	Division Ready Brigade
DRU	Direct Reporting Unit
DTMS	Digital Training Management System
DTTP	doctrine, tactics, techniques, and procedures

EMM	Event Menu Matrix
ERP	Enterprise Resource Planning
FBCB2	Force XXI Battle Command, Brigade and Below
FCS	Future Combat Systems
FCX	fire coordination exercises
FOA	Field Operating Activity
FORSCOM	U.S. Army Forces Command
FTX	field training exercises
FY	fiscal year
FYDP	Future Years Defense Program
GIG	Global Information Grid
GOSC	General Officer Steering Committee
HQDA	Headquarters of the Department of the Army
HRC	Human Resources Command
HS	home station
ICH	instructor contact hours
IDEF0	Integrated Definition for Functional Modeling
IED	improvised explosive device
IMCOM	Installation Management Command
IMI	Interactive Multimedia Instruction
IMT	Initial Military Training
INCOPD	Institute of NCO Professional Development
ISR	intelligence, surveillance, and reconnaissance
IT	information technology
ITRM	Individual Training Resource Model
KM	knowledge management
LD&E	CAC Leader Development and Education
LMS	Learning Management System
LVC	live, virtual, and constructive
M&RA	Manpower and Reserve Affairs
MCTP	Mission Command Training Program
MDEP	Management Decision Evaluation Package
METL	mission-essential task list
MER	Mission Essential Requirements
MOS	Military Occupational Specialty
MOUT	military operations on urban terrain
MSG	Master Sergeant
MTT	Mobile Training Team

NCO	noncommissioned officer
NCOA	Noncommissioned Officer Academy
NCOES	Noncommissioned Officer Education System
NETCOM	Network Enterprise Technology Command
O&M	Operations and Maintenance
OEF	Operation Enduring Freedom
OES	Officer Education System
OIF	Operation Iraqi Freedom
OIL	observations, insights, and lessons
OPFOR	opposing forces
OPORD	Operations Order
OPTEMPO	operational tempo
OSD	Office of the Secretary of Defense
PA&E	Program, Analysis, and Evaluation
PAF	Prepare the Army Forum
PCS	Permanent Change of Station
PEG	Program Evaluation Group
PEO STRI	Program Execution Office for Simulations, Training, and Instrumentation
PERSTEMPO	personnel tempo
PII	personally identifiable information
PME	Professional Military Education
POI	Program of Instruction
POM	Program Objective Memorandum
POM/BES	POM/Budget Estimate Submission
PPBES	Planning, Programming, Budgeting, and Execution System
QLDR	Quarterly Leader Development Review
RC	Reserve Component
SAMS-E	Standard Army Maintenance System—Enhanced
SAT	Systems Approach to Training
SFC	sergeant first class
SGM	Sergeant Major
SL	Skill Level
SLC	Senior Leader Course
SMA	Sergeant Major of the Army
SMDR	Structured Manning Decision Review
SOA	service-oriented architecture
SRG	Senior Review Group
SSG	Staff Sergeant

STRAC	Standards in Training Commission
TADLP	The Army Distributed Learning Program
TADSS	training aids, devices, simulations, and simulators
TADT	The Army Distance Learning Program
TADV	Training Development
TAG	The Adjutant General
TAMIS	Training Ammunition Management Information System
TAPDB	The Army Personnel Database
TCCW	Training Coordination Council Workshop
TCM	TRADOC Capability Managers
TD	Training Development
TD2	Training Doctrine and Development
TDY	temporary duty
TESS	Tactical Engagement Simulation Systems
TGOSC	Training General Officer Steering Committee
THP	take-home package
TRAC	TRADOC Analysis Center
TRADOC	U.S. Army Training and Doctrine Command
TRAP	Training Resource Arbitration Panel
TRM	Training Resource Model
TRS	Training Support
TSGT	T-Sergeant
TSP	Training Support Package
TSS	Training Support System
TTPEG	Training Program Evaluation Group
UARS	Unit Automated Reservation System
UFR	unfunded requirement
USACE	U.S. Army Corps of Engineers
USAR	U.S. Army Reserve
USARC	U.S. Army Reserve Command
USASMA	U.S. Army Sergeants Major Academy
USR	Unit Status Report
VTT	video tele-training
WOES	Warrant Officer Education System

CHAPTER ONE

Introduction

Background

Trained units and competent leaders have always been, and remain, critical elements of Army operational success. The Army Training and Leader Development (ATLD) system, as outlined in AR 350-1, *Army Training and Leader Development,* is the way it develops these individual and collective competencies.

Training and leader development are different but related functions. Training is defined as

an organized, structured process based on sound principles of learning designed to increase the capability of individuals or units to perform specified tasks or skills. Training increases the ability to perform in known situations with emphasis on competency, physical and mental skills, knowledge and concepts.

Leader development is defined as

the deliberate, continuous, sequential and progressive process, grounded in Army values that grows Soldiers and civilians into competent and confident leaders capable of decisive action. Leader development is achieved through the life-long synthesis of the knowledge, skills, and experiences gained through the developmental domains of institutional training and education, operational assignments, and self-development.[1]

As illustrated in Figure 1.1, the mutually supporting ATLD system outputs are trained units and competent leaders. Both of these outputs support the ultimate objective of near- and long-term operational readiness.

From an Army enterprise perspective, the customers are operational commanders, both units and the Army Service Component Commands (ASCC) that provide the Army forces supporting geographic combatant commands (COCOMs). The Army's process for managing ATLD programs must take into account and respond to their needs. Moreover, given the quickly changing nature of today's operational environment and uncertainty of future operational requirements, the ATLD management process must have the flexibility to adapt its programs quickly. Finally, training resources are limited, and the ATLD process must be able to make the best possible use of available resources.

Six primary ATLD activities work in concert to have a direct role in achieving ATLD outputs:[2]

---

[1]   Both definitions are from HQDA, AR 350-1, *Army Training and Leader Development,* December 2009.

[2]   The definitions are paraphrased from HQDA, AR350-1, *Army Training and Leader Development,* December 2009.

Figure 1.1
Training and Leader Development Strategies, Primary Activities, and Outputs

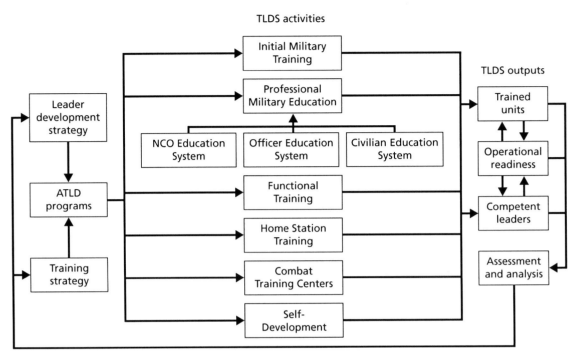

RAND *TR1236-1.1*

- Initial Military Training (IMT) (top of Figure 1.1) provides an orderly transition from civilian to military life. It teaches soldiers and officers the tasks, supporting skills and knowledge to be proficient at the first unit of assignment. IMT includes initial mandatory training to be qualified in a Military Occupational Specialty (MOS) or branch.
- Professional Military Education (PME) are structured leader educational courses conducted by proponent schools and centers for noncommissioned officers (NCOs), officers and warrant officers.
- Functional Training courses are designed to train leaders, soldiers, and Department of the Army (DA) civilians for assignment to duty positions that require specific functional tasks, skills, and knowledge.
- Home Station (HS) Training involves individual, leader, and collective training done at unit home stations.
- Combat Training Center (CTC) Training involves major collective training exercises for brigade and above units conducted at a set of well-resourced training centers.
- Self-development involves all the learning activities done by the individual soldier, with the guidance and support of the soldier's chain of command, for self-improvement.[3]

Each primary activity has a large number of subordinate and supporting activities.

---

[3]   While self-development is obviously an important activity, it is far less structured than the other primary ATLD activities.

The ATLD strategies involve determining how these activities work together, what they should achieve in terms of supporting overall ATLD goals, and how resources should be allocated to them to achieve the highest possible benefit. ATLD strategies are complex both in terms of their means and outputs, and also in terms of the management processes that support them.

The Army's Training and Leader Development Programs Have a Difficult Task

Changing requirements add to the complexity of ATLD management. The range of tasks and skills at which soldiers, leaders, and units must be proficient has increased over the past decade. Traditionally, ATLD programs could focus primarily on the defeat of conventional enemy forces. This mission remains, but operations in Afghanistan and Iraq have demonstrated the need to defeat unconventional forces, protect and support local populations, and perform some civil functions. The requirement for the Army to be capable of conducting operations across a broad range of possible missions and operational environments is codified in Army Doctrinal Publication (ADP) 3-0, *Unified Land Operations*.[4]

Preparing units and leaders to execute the full spectrum of potential missions is, and will remain, a very difficult task for the Army, especially in an era of significant ongoing operations. It will require a thoughtful balance between likelihood and risk.

While ATLD requirements for full-spectrum operations have increased significantly, there is no reasonable prospect for increased funding for ATLD programs. Indeed, future reductions are fairly certain.[5] Moreover, the time available for unit and institutional training and leader development programs, because of continuing operational demands for the foreseeable future, will remain a constraint. This combination of increased requirements and constrained resources has generated a need to make major changes in ATLD strategies and programs.

Institutional Adaptation

A similar need for significant change occurs across the full range of Army programs, and, to support achieving the level of change needed, the Army has undertaken an Institutional Adaptation initiative.[6] The concept is to make decisions for the greatest benefit of the Army as a whole—the Army "Enterprise."

Institutional Adaptation comprises three elements:

---

[4]   This is an Army Capstone Manual. Dated October 2011, it was published after the research for this report was completed, but the report is consistent with its concepts. ADP 3-0 has a stated purpose of providing "a common operational concept for a future in which Army forces must be prepared to operate across the range of military operations, integrating their actions with joint, interagency, and multinational partners as a part of a larger effort." Thus the need for operating across the full spectrum of conflict remains a requirement in this edition—although use of the term full-spectrum operations is less explicit. In this report, we use the term "full-spectrum" operations to mean that future operations can take place across a wide range of potential environments; that operations can have a mix of offensive, defense, and stability components; and that these operations are likely to be highly complex.

[5]   When the research for this project was first done, the likelihood for increased ATLD program resources was low. As of late 2012, budgetary outlooks reduce these chances to the point where even limited ATLD reductions would be considered a relatively favorable outcome for the ATLD community.

[6]   Institutional Adaptation, and its elements, was directed by a January 2009 Secretary of the Army/Chief of Staff of the Army Memorandum, *Institutional Adaptation and Transformation*. Since the research for this report was completed, this initiative has continued to evolve with limited additional formalization. But its goals remain reasonable, and we use them as a framework assessing the Army's system for managing ATLD programs.

- better supporting the Army's Force Generation (ARFORGEN) process for providing ready forces to meet operational requirements[7]
- adopting an Enterprise Approach by developing an Army-wide strategic management system that incorporates a refined governance process supported by an improved assessment architecture
- reforming requirements and resourcing processes by establishing a more responsive and realistic process and promoting good resource stewardship.

To support Institutional Adaptation, the Army formed four Core Enterprises (CE) that are functionally aligned to the Secretary of the Army's Title 10 responsibilities.[8] These entities establish a forum for collaboration among the Army's senior leaders and other stakeholders to share ideas and propose solutions to common problems. The Secretary of the Army and Chief of Staff of the Army (CSA) also established the Army Enterprise Board (AEB) to synchronize issues across functional areas.[9] The four CEs are:

- **Human Capital,** chaired by the Commanding General U.S. Army Training and Doctrine Command (TRADOC) and overseen by the Assistant Secretary of the Army for Manpower and Reserve Affairs (ASA M&RA)
- **Materiel,** chaired by the Commanding General U.S. Army Materiel Command (AMC) and overseen by the Assistant Secretary of the Army for Acquisition, Logistics, and Technology
- **Readiness,** chaired by the Commanding General U.S. Army Forces Command (FORSCOM) and overseen by the Under Secretary of the Army. In the role as chair of the Readiness Core Enterprise, the Commanding General FORSCOM represents the operational force.
- **Services and Infrastructure,** chaired by the Commanding General U.S. Army Installation Management Command (IMCOM) and overseen by the Assistant Secretary of the Army for Installations and Environment.

Institutional Adaptation requires the institutional Army to re-examine its processes at a fundamental level and to make whatever changes are required to provide better support for the operational force. Identifying needed changes and directions for improvement poses a major challenge in an area as complex as the numerous ATLD programs.

---

[7] ARFORGEN is the Army's process of managing the preparation and deployment of ready forces to support operational requirements. Thus support of ARFORGEN processes means focusing the efforts of the institutional Army on operational readiness—supporting COCOM and ASCC commanders' operational requirements, and supporting the unit readiness levels needed to meet these requirements. The ARFORGEN process is described in detail in Chapter Two.

[8] There is no formal documentation of the roles, responsibilities, and authorities of Core Enterprise forums in HQDA, AR10-87, *Army Commands, Army Service Support Commands, and Direct Reporting Units*, dated September 2007. This regulation "prescribes the missions, functions, and command and staff relationships with higher, collateral headquarters, theater-level support commands, and agencies in the Department of the Army (DA) for Army Commands (ACOMs), Army Service Component Commands (ASCCs), and Direct Reporting Units (DRUs)." Thus the specifics of Institutional Adaptation are under debate and still emerging. However, implementation guidance is being transmitted informally and implementation has started. See DA briefing, HQDA, *Institutional Adaptation*, November 2009.

[9] See HQDA, *Army Enterprise Board Charter*, May 2009. The CSA presides over the AEB, and its membership includes the DA and Army Command senior leaders. Its "purpose is to advise the Secretary of the Army" and "it serves as a forum for collaboration and synchronization within the Department of the Army."

Purpose

Many in the Army's training and leader development community were concerned about their ability to make the case for the resources needed to support their programs under the Institutional Adaptation construct. In addition, they believed that presenting objective analysis to show the relationship between ATLD program resources and readiness to support balanced resource decisions among training, manning, and equipping functions, which had been a historic challenge, would become even more important.[10] Given these concerns, the DA Director of Training and the U.S. Army Combined Arms Training Center's Deputy Commander-Training asked RAND's Arroyo Center to support their institutional adaptation efforts. The purpose of this study is to support efforts to adapt the ATLD program's requirements and resourcing processes by applying an enterprise approach and sharpening the focus on ARFORGEN process support. Its primary objective is to identify important directions the Army can take to improve DA level ATLD management processes and architectures to better support operational readiness and effective stewardship of resources. While the major research effort for this report was completed in 2010, follow-on research and coordination shows that the major findings, conclusions, and suggested directions with regard to adapting ATLD management processes remain valid and relevant.[11]

Assumptions

We make three assumptions. First, that the Army will continue to deploy forces to support actual operations, but the number of units that deploy will be significantly less than was the case in 2011, and many units in an ARFORGEN cycle will not deploy at all.

Second, that the Army will face a wide range of potential operational requirements, and the Army will have to train units and develop leaders on a wide range of combat and non-combat tasks to meet these requirements.

Finally, we assume there will be increased pressure to reduce ATLD budgets and that efficient use of ATLD resources will be increasingly important.

Approach

Our approach involved the use of three previous ATLD-related studies and a new case study effort of a specific ATLD activity. As described above, a large number of different activities and a large number of Army organizations are involved in ATLD's management and execution. For this reason, an examination of a wide range of ATLD activities and management processes was necessary to develop valid conclusions about overall ATLD process improvement.

Two of the previous efforts looked at operational force training. The first supported the Army's efforts to select the most important training enablers for modernized Brigade Combat

---

[10] See Defense Science Board Task Force, *Training for Future Conflicts*, June 2003; and Army Science Board, *Technical and Tactical Opportunities for Revolutionary Advances in Rapidly Deployable Joint Ground Forces in the 2015–2025 Era*, Summer 2000.

[11] The Army is a dynamic institution. Terms, concepts, regulations, and many other aspects are constantly changing. To the degree possible, this report has been updated.

Teams (BCTs), and the second examined the Army's Training Support System (TSS) processes for providing key resources to support training and leader development. A third examined the Army's Distributed Learning (DL) program—which is designed to enable the delivery of training and learning to soldiers and leaders wherever and whenever they need it.

In the case study, we examined involved training and leader development in a specific TRADOC PME course—the Advanced Leader Course (ALC) for junior NCOs. The case study supplemented the previous research in that it examined institutional training and had a more direct focus on strategic management processes. In the case study, we examined ALC's "As-Is" management processes and architectures. We did so in sufficient detail to understand what key decisions must be made and what information is both needed and available to support strategic decisionmaking. Even though the focus is on DA-level management, the training activity is examined from the execution level to DA level.

We used a research-based qualitative analysis approach. We first examined the ongoing ATLD processes to understand how they function and to identify the organizations that participate in the processes and their roles. We started by examining relevant Army regulations (AR), policies, pamphlets, and other documentation outlining ATLD processes and outputs. However, ATLD processes constantly evolve and much of the formal documentation, such as in regulations, is not current. Therefore, we reviewed a wide range of other documentation, such as briefings and concept papers. In addition, a central element of our research involved discussions with a range of participants in the ATLD community concerning the specifics of current processes and directions being taken.

Our next step was to assess the overall process. We used the following questions to assess the extent to which change to an Institutional Adaptation approach is warranted and being achieved:

- **Support operational readiness and ARFORGEN processes.** To what degree have management processes been adapted to support changing unit operational readiness requirements in the context of the ARFORGEN process and a constrained resource environment?
- **Adopt an enterprise approach.** Do management processes focus on overall ATLD benefit and are they supported by structured assessment architecture?
- **Stewardship of resources.** To what extent are decisions made after a systematic consideration of overall costs and benefits?

The criteria implied by these questions align with the goals of Institutional Adaptation listed above. Answering the questions will necessarily have subjective elements, but is evidence-based.

Based on the answers to these questions, we developed findings and conclusions about the adequacy of ATLD management process. From these, we identified areas for improvement and then developed directions the Army could take to improve its management process and architectures across the wider range of ATLD programs.

Case Study Selection
Three considerations shaped selection of an appropriate training activity for the case study:

- It should complement the previous research.

- It should be broad enough to represent a wide range of ATLD management issues.
- It should be narrow enough for the research team to examine it in detail.

The one selected jointly by the study team and study sponsors was the ALC at Active Component unit home stations. ALC is a Non-Commissioned Officer Education System (NCOES) course that provides general leadership and basic branch-specific, squad- and platoon-level training to mid-grade NCOs.[12] While ALC is a reasonably focused ATLD activity, its execution requires coordinated support from a range of ATLD community members. ALC is an important component of the ATLD program. Since we are examining its execution at unit home stations, execution involves both TRADOC, which provides instructors and courseware, and units, which provide students, equipment, and other course support. Portions of ALC are taught in classrooms as well as in field locations, and it has a DL component.

Organization of This Report

This report discusses our examination of ALC and presents our conclusions regarding achieving the goals of Institutional Adaptation within training and leader development programs. A brief description of each chapter's contents follows:

- Chapter Two describes the role of the ALC in the Army's training and leader development strategies as well as current challenges to ALC.
- Chapter Three presents findings and conclusions from our analysis of ALC management and execution and outlines their implications for Institutional Adaptation of ATLD program management processes.
- Chapter Four summarizes relevant findings and conclusions from previous ATLD program–related research and outlines their implications to Institutional Adaptation of ATLD management processes.
- Chapter Five outlines overall conclusions about areas in which ATLD management could be improved and outlines directions the Army could take to better manage the broad range of ATLD programs.
- Chapter Six summarizes the major conclusions from our research and discusses the broader implications for management of the full range of training and leader development programs.

---

[12] HQDA, AR 350-1, 2009. The specifics of NCOES and ALC are discussed in detail in the next chapter.

CHAPTER TWO

Advanced Leader Course's Role and Challenges

This chapter describes the ALC's role in the Army's training and leader development strategies and its current challenges. It begins by describing the ALC and the benefits it provides different audiences, and then discusses how it fits into the ARFORGEN cycle and the difficulties the cycle creates for getting soldiers to and through ALC. It next describes the steps the Army has taken to transform ALC so that it better meets the needs of the Army and the soldiers who attend it. It concludes with a discussion of other potentially beneficial ALC changes.

ALC and Its Customer Benefits

AR 350-1 describes the role of the overall NCOES program:[1]

> The goal of NCO training and the NCOES is to prepare noncommissioned officers to lead and train soldiers who work and fight under their supervision, and to assist their leaders to execute unit missions. NCOES is linked to promotion to SSG [staff sergeant], SFC [sergeant first class], MSG [master sergeant], and SGM [sergeant major]. This ensures Non-Commissioned Officers (NCOs) have the appropriate skills and knowledge required before assuming the duties and responsibilities of the next higher grade.

Within NCOES, ALC provides "leader training and basic branch-specific, squad- and platoon-level training." It is the first NCOES course that provides training on MOS-related technical and tactical skills, and is a requirement for promotion to sergeant first class E7.[2] This goal represents a recent expansion. ALC replaced the Basic NCO Course (BNCOC) in 2009. BNCOC content was designed to train squad-level leader and MOS-specific tasks and skills only, and previously BNCOC was a requirement for promotion to staff sergeant E6.

Most ALC courses have two phases: a common core phase and MOS-specific phase. The common core phase is conducted online in a DL mode, covers selected squad-level non-MOS–specific tasks and skills, and is designed to take about two weeks to complete.[3] The ALC Common Core was developed and is taught by the U.S Sergeants Major Academy at Fort Bliss.

---

[1] AR 350-1.

[2] AR 350-1, p. 77.

[3] While the course has been redesigned to be delivered in a Web-based version, training and learning objectives for this course are outlined in U.S. Sergeants Major Academy, *Basic Non-Commissioned Course Program of Instruction*, June 2008, and remain unchanged.

The second phase is MOS-specific and can be up to eight weeks long.[4] This phase is done face-to-face either in residence or by having Noncommissioned Officer Academy (NCOA) instructors conduct the course at unit installations by Mobile Training Teams (MTT). When done in residence, this phase of ALC is done on a temporary duty (TDY) basis, either between permanent change of stations or while the soldier is assigned to a unit. There also are options for having an MOS-specific DL phase in addition to the live phase, but this option is exercised for only a small number of MOSs. Thus, the time for taking the face-to-face portion generally comes out of the unit commander's time for training or other unit activities, because it is time in which the NCO is away from the unit.[5]

Courses of instruction for the MOS phases of ALC are developed by proponent schools, and, for the active component (AC), taught by NCO academies that are located at the proponent school or center and that are under the command of the proponent's commanding general.

NCOES completion is required for promotion, with ALC completion being required for promotion to sergeant first class/platoon sergeant (E7). But in certain cases, waivers, such as for a deployment, can be granted and the soldier is conditionally promoted.[6]

ALC has both costs and benefits for Army customers. These are outlined in Table 2.1.

ALC has three customers: the soldier taking the course, the chain of command to which the soldier belongs, and the Army as an enterprise. The costs are, to the soldier, time away from family and job; to the unit commander, the loss of the soldier for the period the soldier is attending the course and any support that must be provided to execute an MTT; and to the Army, the dollar, manpower, and other resources needed to support execution. The benefit to all customers is an NCO who is better trained in the requirements of the position and one who possess greater leadership skills. This relates to increased unit readiness in the near and long term. Perhaps a more immediate and important benefit for many soldiers is that completion of ALC is a requirement for promotion.

Table 2.1
ALC Customer Costs and Benefits

Customer	Costs	Benefits
Soldier	• Time away from family/job	• Near- and long-term job skills • Promotion
Unit Chain of Command	• Soldier away from unit • Support of course (MTT)	• More capable leaders/unit • Take care of soldier (promotion)
Army Enterprise	• Program Objective Memorandum (POM) $ • Military Manpower	• Unit readiness • Individual leader development

---

[4]   There are exceptions, and some courses for highly technical MOSs, for example, some intelligence and medical MOSs, are longer.

[5]   By contrast, Officer Education System courses are done on a Permanent Change of Station (PCS) basis in which the officer is assigned to the school and not to a unit.

[6]   See HQDA, AR 600-8-19, *Enlisted Promotions and Reductions*, dated March 2008, and DA G-1 Memorandum, *Non-Commissioned Officer Education System (NCOES) Deferral Policy*, March 2008. If conditionally promoted, the soldier must complete the required course within 270 days of return.

Stresses on ALC

The current operating environment and increased operational training and leader development requirements have posed significant challenges for those responsible for achieving ALC training and leader development goals.

Increased Requirements Without Longer Courses

The requirements for NCOES and unit training programs have greatly increased. In the contemporary operating environment, Army units and their leaders must prepare for a wide range of possible operational missions under an even wider range of possible conditions. An obvious result is that the range of tasks and skills expected of the Army NCO has grown dramatically. Among the skills frequently central to full-spectrum mission accomplishment are cultural awareness, an ability to work with local and interagency governmental personnel, an ability to work through an interpreter, and many other similar skills.

A second factor that increases the requirements of ALC is the expansion of course objectives. The predecessor BNCOC had the objective of training squad leader skills, but ALC has the objective of training both squad leader *and* platoon sergeant skills—a major increase. While the course objectives and NCO skills requirements have increased, guidance has been to maintain or reduce current course lengths.

Persistent Operational Deployments

Before 2002, while unit requirements for training and other requirements were large, there were reasonably large windows in which unit schedules could accommodate sending NCOs to NCOES. That is no longer the case. A large percentage of Army units are deployed, and the others either are recovering from deployment or preparing to deploy again.

To meet current and future operational demands, the Army has developed and implemented a process, ARFORGEN. ARFORGEN is an approach to synchronizing requirements in a logical, systematic manner to provide regional commanders the range of full-spectrum—capable forces needed to meet ongoing and contingency requirements.[7] The goal is to provide a continuous output of modernized expeditionary forces to meet known and planned COCOM operational requirements, and to handle any unplanned operational requirements as well. Training and leader development strategies and programs support ARFORGEN processes by meeting immediate and long-term COCOM/ASCC requirements for trained units and developed leaders.

Under ARFORGEN, units cycle through three pools, each connoting a higher state of readiness and availability for operations: Reset, Train-Ready, and Available:[8]

- **Reset**: Units start this phase upon return from deployment or after one year in the Available pool. The intent of Reset is to restore the unit's personnel and equipment to a deployable level at the end of six months (12 months for the Reserve Components) so they can begin preparing for their next mission.

---

[7]   Headquarters, Department of the Army, *Army Campaign Plan, Change 5*, April 2007.

[8]   Memo CSA, *Army Training and Leader Development Guidance* FY10-11, July 2009.

- **Train-Ready**: Collective training in this phase achieves full-spectrum readiness. As specific operational missions are assigned, units train against expected conditions and perform preparations for deployment and in-theater operations.
- **Available**: Units in this phase are fully prepared for rapid deployment to meet planned or unplanned contingencies. Some will deploy, and some will remain in deployable status and will be the first deployed if forces are needed.

Under the ARFORGEN concept, a goal is that NCOES and other PME courses be accomplished while units are in the Reset pool. This is to ensure that NCOs have the skills needed to support collective training during the Train-Ready period.

The ARFORGEN process and the role of ALC in supporting it are shown in Figure 2.1.[9]

The length of the Reset period provides only a narrow window for NCOs to attend ALC. These courses generally last up to eight weeks, and many important activities take place during the six-month Reset period. A key activity is soldier recovery. The intent for Reset is to rest recently deployed soldiers, and travel away from home station is to be strictly limited. The current policy is that soldiers will not be scheduled for NCOES until 90 days after their return from deployment, and scheduling attendance at an eight-week resident course in the remaining 90 days of the Reset phase is not always possible. Scheduling attendance during Reset is difficult given that during this period, sending soldiers to NCOES is only one among many important activities that must be accomplished.

Figure 2.1
ARFORGEN Process and How ALC Supports It

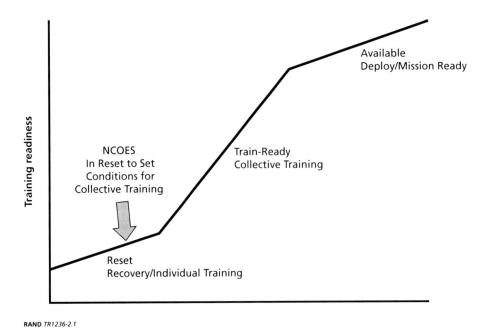

---

[9]   HQDA, *Army Campaign Plan*, April 2007.

As a result of the current operational pace, almost one-third of AC staff sergeants have not attended ALC.[10] This backlog further complicates scheduling and undercuts unit readiness, because it means that many unit leaders will not have formal training in important tactical, technical, and leadership skills.

The range of experience and competence for prospective course attendees varies considerably. Some junior soldiers will be preparing for assignment to staff sergeant positions, while others will have served in such positions for many years. To illustrate this range of experience, we examined the time in service of AC NCOs from three separate MOSs who attended BNCOC in 2008. About half had between six and ten years' service; about a quarter had less than six; and about a quarter had more than ten years' service. The grades ranged between sergeant and sergeant first class, with more than half already being staff sergeants—the grade the course was designed to teach. This range of student experience is another factor that makes it difficult to shape the course for full individual soldier relevance and benefit.

ALC Transformation

The Army recognized that the nature of changed requirements and rate of unit deployments created a need for more major change and began examining the options. In 2005, the commanding general of TRADOC directed TRADOC schools and centers to conduct a detailed analysis of their respective Advanced NCO Course (ANCOC) courses to "transform the structure and content of NCOES to support an Army at War, the modular force, stabilization and the ARFORGEN model."[11] A revised set of courses started in FY 2010. Under NCOES transformation, the BNCOC course was changed to the ALC. Key elements of NCOES Transformation in regards to ALC included:

- The use of MTT to perform MOS-specific ALC course phases—allowing attendance at the soldier's HS rather than TDY to the resident school.
- Reduction in the length of many longer ALC courses to eight weeks, and in some cases less.
- As discussed earlier, the course goals were expanded to include both staff sergeant and sergeant first class skills.
- The online DL ALC Common Core course was developed and fielded.

The NCOES transformation initiatives appear to have reasonable benefits. In particular, schools, students, and commanders have viewed the use of MTTs as a major success. It has reduced soldiers' time away from home during Reset and made scheduling attendance easier and less burdensome on commanders.[12] Reducing course length has similar benefits.

---

[10] Data from the DA G-1 show that as of August 2008, 19,633 of 64,416 staff sergeants had not graduated from BNCOC. Our discussions with the DA G-3's leader development staff indicate that the size of this backlog had not declined by the end of FY 2009.

[11] Combined Arms Center, Headquarters, *Operations Order 05-165A, NCOES Transformation*, July 2005.

[12] John C. Morey et al., *Best Practices for Using Mobile Training Teams to Deliver Noncommissioned Officer Education Courses*, ARI Report, No. A943005, 2009. Discussions with ALC cadre and students during our visits supported this report's finding.

Including sergeant first class skills in ALC also makes sense, because many of the students are experienced staff sergeants and in some cases even sergeants first class, so this could add to course relevancy. It also aligns the course to the promotion requirement. The primary motivation to attend ALC is promotion, and as long as an ALC is a requirement for promotion to sergeant first class, a large percentage of attendees will be senior staff sergeants.

Moving to a Web-based delivery of some common core content increases flexibility. Previously, this portion of the course was taught either in residence or by video tele-training (VTT), which either required travel time or created facility availability issues.

Further Change Is Possible and Could Increase Benefits

However, further change to provide better support to soldiers and leaders appears to be both feasible and potentially beneficial, especially in an environment in which frequent deployments constrain possible course length and complicate attendance.

The addition of the MTT option has had benefit, and this is likely a key reason the backlog of ALC has remained stable. But there is limited possibility to expand this option to better align with the ARFORGEN cycle, support the Army's goals of keeping soldiers at HS during Reset, and reduce the backlog. Almost all of the MOSs with a large enough student population to make an MTT approach possible already are using this approach. But the majority of NCOs are in MOSs in which the annual ALC training requirements are less than 200. These small numbers make it generally impossible to find a population at any single post in a Reset window large enough to make an MTT approach practical.

Greater use of DL approaches has the potential of improving timely attendance in the context of an ARFORGEN cycle, allowing the soldier to get ALC instruction during Reset with reduced time away from home station. But there appears to have been almost no movement in this direction.

The recent reshaping of the Special Forces ANCOC illustrates this potential for greater use of DL. The resident phase of this course was reduced from seven and a half weeks to three weeks through the use of a combination of computer-based Interactive Multimedia Instruction (IMI) and a collaborative DL phase. In the collaborative portion of the DL phase, instructors and students interacted online asynchronously, providing many of the benefits of classroom interaction.[13] Such blended use of stand-alone and collaborative DL and resident instruction is becoming increasingly prevalent in business and academia, and the use of "blended DL" allows more flexibility in starting and conducting the course at times suited to student and unit needs. We reviewed several ALC Programs of Instruction (POIs) and found that all had much—and in some cases most—of the material that was taught in a classroom. The increasing use of DL by both civilian and military academic institutions indicates that much of the material that is taught in a classroom can also be taught using DL methodologies with little or no drop-offs in learning.

Improvement in supporting the achievement of overall ALC goals also could occur by providing options for tailoring material to individual student needs to take into account the wide ranges in student experience. ALC students range from sergeants to sergeants first class.

---

[13] The effort and the benefits of a blended DL (one using both synchronous and asynchronous DL) approach are described in more detail in Michael G. Shanley, James C. Crowley, Matthew W. Lewis, Susan G. Straus, Kristin J. Leuschner, and John Coombs, *Making Improvements to the Army's Distributed Learning Program*, Santa Monica, Calif.: RAND Corporation, MG-1016-A, March 2012. In terms of interaction, during interviews, Special Forces ANCOC course instructors told us that the ability to interact with, and among, students was actually greater in a collaborative DL environment than in a classroom.

That means three grade levels and around ten years' difference in service separate the most junior and most senior students. The current methodologies have limited to no flexibility to tailor course content to student needs and experience; basically, all students go through the same course. We found no use of pre-tests to allow students either to shorten their time in the course or, in the case of more experienced students, to add subject matter. Again, blended training methods could support such approaches.

Another area in which further change could increase course benefit to the student and the student's unit would be through achieving better integration between the course learning objectives with unit training programs during the ARFORGEN cycle to better support the needs of operational force commanders. Shaping the course training objectives to include those that would best support train-the-trainer and tactical skills needed to support the collective training programs. Again, tracking the students would provide greater benefit; for example, providing a different track for an NCO going into a squad leader position from that provided to an NCO assigned to a platoon sergeant position. Ideally, the course would cover the full range of critical leader tasks and skills. But the course lengths for most MOSs are not long enough for this, especially considering the increase in course goals to include both Skill Level 3 (staff sergeant) and Skill Level 4 (sergeant first class) tasks in the same course.

The course also could be focused on the equipment in the NCO's unit. An argument can be made for training on the full range of equipment required by the MOS. But again, the shortened course lengths and the consideration of skill decay for technical training that will not be reinforced subsequent to the course suggest that the benefit from training on equipment that is not in the unit is limited at best. The sharper focus will, logically, better support unit readiness.

A final consideration is the alignment of ALC resources with training objectives. The addition of sergeant first class skills expands course goals, but there has been no increase in course time. In fact, the emphasis has been on reduction, so the amount of time available for each skill level being taught has been effectively reduced. Also, the instructor grades have not been increased to provide for instructors with sergeant first class experience to teach the elevated skill sets.

These potential changes are not presented for the purpose of recommending specific ALC changes—each would involve major change and require further study. The point is that there are feasible and potentially beneficial ALC changes that should be considered. This study examines the potential for improving the ability of the ATLD management process to make beneficial changes; this is an issue we examine in the next chapter.

CHAPTER THREE

ALC Management Findings and Conclusions

In this chapter, we examine the processes for management and execution of ALC and present the findings and conclusions we drew from this examination. While we focus on Army-level decisions and the information needed to support making them, we do this from the perspective of execution of ALC instruction with regard to effective use of resources and benefit to operational force readiness.[1]

ALC Management Processes

ALC management processes are complex. To organize our effort, we first identified the major management activities in this process and then examined and developed findings and conclusions for each.

We identified four levels of ALC management activities. These levels are Strategic, POI and Courseware Development, Program, and Execution. We discuss each in turn, and indicate for each where the information to support the process is available:

- **Strategic.** Management decisions at this level focus on overall goals and resource allocations and integration of ALC into broader training and leader development strategies. It also includes oversight and adjustment.
- **POI and Courseware Development.** Design ALC courses to achieve goals and objectives within the resources allocated.[2]
- **Program Management.** Implement and support ALC training to achieve course goals, objectives, and learning outcomes.
- **Execution.** Execute the courses to achieve learning outcomes.

These management and execution activities are supported by information support systems and information technology. Determining whether the information needed for effective management and execution was available and accessible is a fifth key ALC management aspect we examined.

---

[1] This examination was completed in February 2009. As with most management processes, these are under constant revision. However, in 2011, we verified that, while some changes had been made, the major findings and conclusions presented in this chapter remain valid.

[2] By "courseware" we mean the formal documentation of specifically what is taught in courses, and how subject matter is taught. In this regard, the key elements of courseware are Training Support Packages (TSPs).

Based on the examinations of these activities and of the information support systems and technologies that support them, we developed the overall ALC management findings and conclusions that are presented at the end of this chapter.

The activities and their relationships are portrayed in Figure 3.1. The box in the upper left of the figure represents the strategic process. The program management process is represented by the box labeled "Manage ALC Training." The POI and courseware development process corresponds to the box in the figure labeled "Revise ALC Content."

This figure, and subsequent ones illustrating management activity process flows, uses the Integrated Definition for Functional Modeling (IDEF0) notation. This technique is widely used in the Department of Defense (DoD) and is relatively easy to use and understand, delivering a fair amount of meaning in a simple graphical form.

An IDEF0 diagram consists of one or more "activities" (shown as rectangular boxes):

- **Inputs** enter an activity from the left and are items or information processed or transformed by the activity, serving as the raw material (if any) for its Outputs.
- **Controls** enter an activity from above and guide, direct, constrain, or provide context for the activity's behavior.
- **Mechanisms** enter from the bottom of an activity and represent techniques, systems, data, or personnel that help an activity do what it does.
- **Outputs** exit from the right of an activity and may become inputs, controls or mechanisms for other activities.

From an IDEF0 perspective, the overall ALC management activity is shown on Figure 3.2.

Figure 3.1
Major ALC Management and Execution Activities

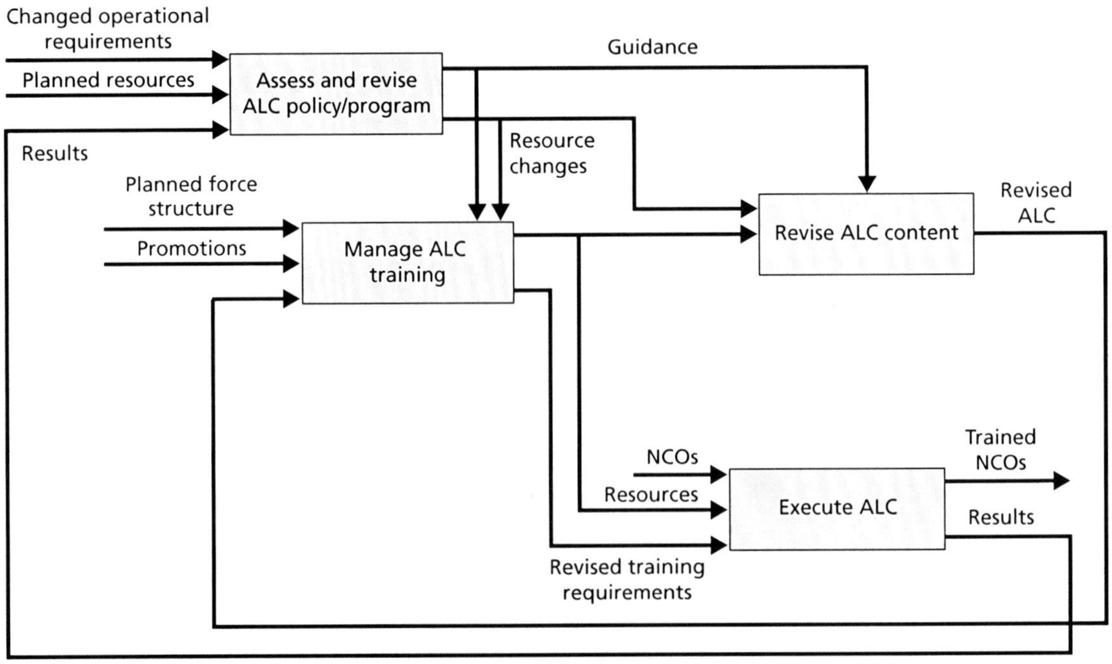

RAND *TR1236-3.1*

Figure 3.2
The ALC Management Activity

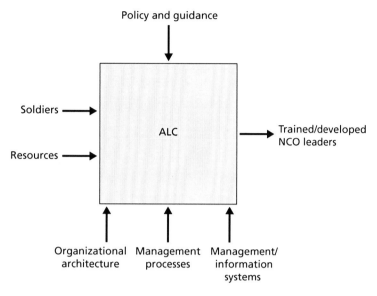

RAND *TR1236-3.2*

ALC takes soldiers and resources as inputs and produces better-trained NCO leaders as its outputs. The primary controls are Army strategic guidance and policies and the primary mechanisms are various organizational architectures, management processes, and management and information systems.

Strategic Management

In this section, we examine the strategic management of ALC and draw conclusions with regards to its ability to achieve the goals of Institutional Adaptation.

Inputs and Outputs. Based on changed requirements, strategic management results in decisions with regard to ALC program revisions, resource reallocations, and other implementation guidance. The IDEF0 view of this activity is illustrated in Figure 3.3.

Strategic Management Requires Difficult Decisions

The Strategic Management activity requires making difficult interrelated decisions to balance and adjust the ALC program to achieve the best possible support of unit readiness and long-term leader development. Several competing considerations complicate decisionmaking:

Length. Only so much time can be allocated to these courses, because the time that the student spends in ALC courses comes out of the limited time that the unit commander has for unit training and other important activities. Likewise, the course length generates costs for student attendance at resident courses or instructor travel to MTT course locations.

Goals and Objectives. ALC has key readiness and leader development roles, and in many ways, its importance is increasing as the range and complexity of needed leader skills increase. But given that course lengths are constrained, the choice among skill levels, basic

Figure 3.3
ALC Strategic Management Process

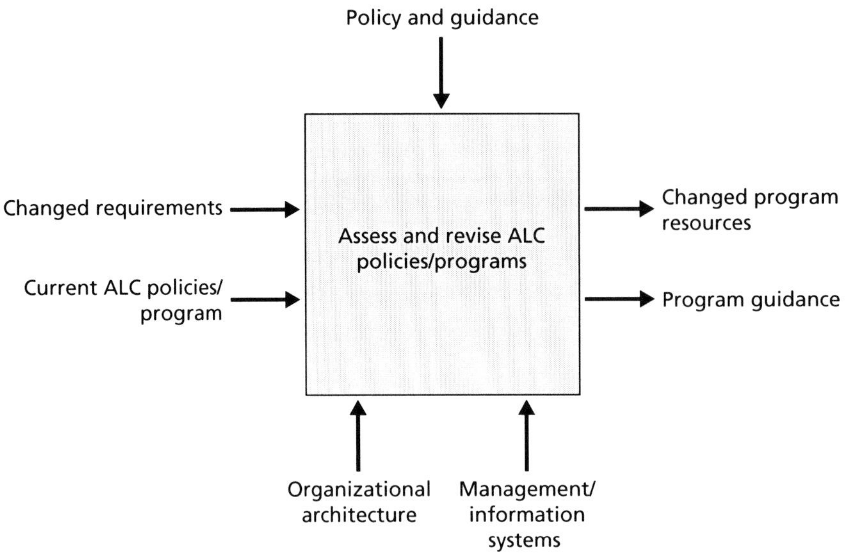

RAND *TR1236-3.3*

leadership, technical skills, and tactical task balance is difficult, with no obvious right answer. Much of it is important, but only a portion can be included.

Relationship to Promotion. This is a very difficult strategic decision. ALC's predecessor, BNCOC, was designed to train critical Skill Level (SL) 3 tasks and leadership skills, and logically would be a requirement for promotion to staff sergeant E6, as used to be the case.[3] But with the current rate of operational deployments, many soldiers have not been able to attend on this schedule. Enforcing an ALC requirement for promotion to staff sergeant E6 would penalize soldiers for an outcome beyond their control, so strategic decisionmakers have to balance training and leader development needs with fairness to soldiers. The current policy is that ALC is a requirement for promotion to sergeant first class E7.

Other Resources. Experienced instructors, constrained facilities, equipment, and training support all have competing unit and other claimants. A decision to support ALC means reduced resources for other priorities.

ATLD strategic decisionmaking involves providing broad areas of policy and program guidance. Most ALC decisions are managed at lower levels based on this strategic guidance. However, given the importance of NCOES, important decisions often are made at the highest levels. For example, there was disagreement between the Army's G1 and G3 concerning the question of "constructive credit"—that is, whether an NCO who had been promoted and served higher than the level for which an NCOES course is designed to teach should still be required to attend the lower course or instead attend the appropriate course for the NCO's grade.[4] The decision not to give NCOs "constructive credit" was made by the CSA.

---

[3]   Skill Levels relate to an enlisted soldier's grade: SL1 to E1–E4, SL2 to E5, SL3 to E6, and SL4 to E7.

[4]   For example, ALC is designed to teach squad sergeant skills and Senior Leader Course (SLC) platoon sergeant skills. If a soldier has been promoted to sergeant first class (and would no longer be assigned to squad leader positions) but has not graduated from ALC, should the NCO be given constructive credit for ALC and instead attend SLC?

Organizational Architectures for Strategic Management Are Complex

The organizational architecture is outlined in Figure 3.4.

As shown in this figure, many organizations are involved in the strategic management of ALC, and each has different perspectives and interests.[5] Thus, guidance to course developers, program managers, and executors can come from many directions, and synchronization is difficult.

ATLD strategic decisionmaking architectures include formal and informal collaboration mechanisms to facilitate decisions being made for the best possible overall Army benefit. Formal collaboration involves forums (indicated on the right side of Figure 3.4) and staffing processes. Informal collaboration involves constant networking among the Army's leadership at all levels.

At the highest level, the strategic management of all Army programs is the responsibility of the **Secretary of Defense** and the **Secretary of the Army**. The **CSA** works closely with the Secretary of the Army on matters of policy and programs. The CSA is responsible for many of the actual management decisions that flow from the policy and other guidance provided by the Secretary of the Army.

Figure 3.4
ALC Strategic Management Organizational Architecture

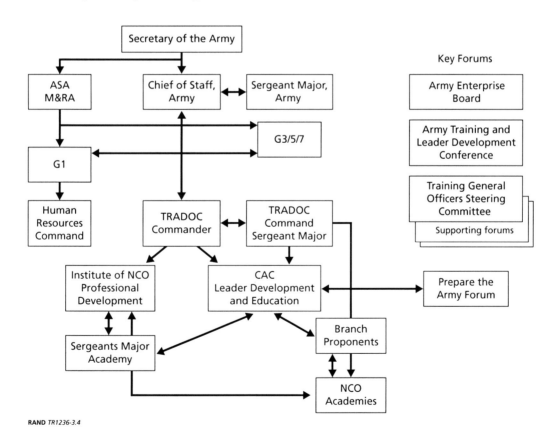

RAND *TR1236-3.4*

---

[5]   The acronyms in this figure represent different organizations and forums, which are defined and described in the next few pages.

The Assistant Secretary of the Army (ASA) for Manpower and Reserve Affairs (M&RA) is responsible for setting strategic direction and providing policy, programming, and oversight of human resources, readiness, and training.[6] The ASA (M&RA) provides oversight for both the G-1 and the G-3/5/7, who also have roles in these areas.

At the strategic level, the Army has established an **Army Enterprise Board (AEB)**.[7] The AEB is a collaborative forum created to advise, assist, and support the Secretary of the Army in making informed decisions that ensure the effective and efficient delivery of trained and ready forces to combatant commanders while preserving the all-volunteer force. The Secretary of the Army presides over the AEB. The Under Secretary of the Army, the Assistant Secretaries of Army, the Chief and Vice Chief of Staff of the Army and the heads of the four Core Enterprises are other AEB members.

The **Army Training and Leader Development Conference (ATLDC)** is a high-level forum that focuses on current and future strategic training and leader development issues. The ATLDC provides an opportunity for informative dialogue among the CSA and senior commanders and Army trainers on changes necessary to support the Army Campaign Plan and Transformation Road Map.[8]

The **Sergeant Major of the Army (SMA)** is the principal advisor to the CSA on enlisted matters. To carry out this responsibility, the SMA traditionally spends extensive time traveling and meeting with soldiers around the world and interacts continuously with an extensive network of senior NCOs throughout the Army. The SMA has a major role in all decisions involving NCOES and other decisions involving enlisted soldiers. No decision in these areas is made without the SMA's review, and, through the CSA, the SMA can generate initiatives in these areas. The Army's NCO "channel" thus has an extensive, positive role in all decisions involving the enlisted force.

The Army **Deputy Chief of Staff (DCS), G-1,** is the Army's personnel proponent and is responsible for "developing, coordinating, and implementing programs and policies directly associated with accession, development, distribution, and sustainment of military and civilian personnel, and the readiness of Army units and organizations."[9] The G-1's responsibilities include determining the broad objectives of the military personnel management system and for supervising the professional development of military personnel to include selection and attendance at Army schools, and development and administration of the Army's military personnel management system, including promotion policies.[10] Thus, many programs and policies regarding ALC fall under the G-1, specifically with regard to ALC attendance, the relationship of ALC to promotions and assignments, and the staffing of ALC academies.

The G-1 also establishes policy and procedures for developing and verifying ALC training requirements. The **Human Resources Command (HRC)** is the G-1's agent for selecting

---

[6]   HQDA, AR 350-1, 2009.

[7]   The AEB was established in mid-2009. Its role has not been formally incorporated into regulations, but this is a description from a DA Enterprise Task Force November 2009 briefing, *Institutional Adaptation.*

[8]   HQDA, AR 350-1, 2009.

[9]   HQDA, *General Orders Number 3, Assignment of Functions and Responsibilities Within Headquarters,* July 2002.

[10]   HQDA, AR 350-1, 2009.

and scheduling soldiers for ALC courses.[11] These processes are described in a later section of this chapter.

The **Deputy Chief of Staff, G-8,** is responsible for integrating Army funding, fielding, and equipping actions with the Office of the Secretary of Defense, Joint, and Army Staff organizations and processes to meet current and future force requirements. The G-8 is also responsible for the development and defense of the Army's planned Program Objective Memorandum (POM) and Future Years Defense Program (FYDP) budgets.[12]

The **Deputy Chief of Staff, G-3/5/7,** is responsible for developing and coordinating policy, programs, and initiatives to achieve directed levels of individual, leader, and unit training readiness for the Army. The DA G-3/5/7 has a Director of Training (DOT) who heads the staff organization supporting the DA G-3/5/7s training and leader development roles. The G-3/5/7 also serves "as the focal point for prioritization, integration, and synchronization of decisions," and thus the DA G-3/5/7 plays a key role in all resource prioritization processes.[13] In addition, the DA G-3/5/7 writes the Army's training strategy. It is also the action office for developing the overall training and leader development policies contained in AR 350-1, *Army Training and Leader Development*; and for developing the CSA's training and leader development guidance, which is published annually.

Decisions on ALC resources are made at multiple levels, but it is at the DA level that the major decisions are made about factors that drive resourcing levels, such as course length. DA also makes overall resourcing decisions with regards to budget allocations to NCOAs and to proponent schools for such functions as training development and DL courseware.

To coordinate training and leader development policies and program direction, the DA G-3/5/7 runs a set of decision support forums involving training, with the primary one being the **Training General Officer Steering Committee (TGOSC)**. The TGOSC provides a management process to identify and resolve issues, determine priorities, and make decisions in support of ATLD.[14] It also develops recommendations for "improvements in training policy and strategy, and capabilities needed to provide trained and ready soldiers, leaders, Army civilians and units."[15]

The TGOSC is chaired by the G-3/5/7. Its primary members are the general officers from the G-1, each ACOM, ASCC, and DRU. General officers from the Army National Guard (ARNG) and U.S. Army Reserve (USAR) are also members. The TGOSC is supported by a number of functional Councils of Colonels and an Integration forum.

The DA G-3/5/7's DOT develops the detailed training and leader development program budgets for CSA approval. The processes of budget development are covered in a later section of this chapter.

---

[11] AR 350-10, *Management of Individual Army Training Requirements and Resources,* September 2009.

[12] General Orders Number 3, *Assignment of Functions and Responsibilities Within Headquarters*, Department of the Army, July 2002.

[13] General Orders Number 3.

[14] HQDA, AR 350-1, 2009.

[15] HQDA, AR 350-1, 2009.

TRADOC develops ALC POIs and courseware, and conducts courses for the AC. The **TRADOC commander** has been designated as the single responsible individual to direct the execution of the Army Leader Development. This program contains all Headquarters, Department of the Army (HQDA)–approved initiatives and provides the management process for program execution, approval of new initiatives, and prioritization.[16]

The **TRADOC Command Sergeant Major** (CSM) has a more formal role in NCOES than the SMA. Besides serving as the TRADOC commander's advisor on all enlisted matters, the TRADOC CSM provides direction and oversight of the NCOES across the Army; provides direction to the Institute for NCO Professional Development (INCOPD) and the U.S. Army Sergeants Majors Academy (USASMA) on NCO development priorities, policies and programs; and serves as the NCO subject matter expert for the Army Leader Development Enterprise.[17]

The **Commander of TRADOC** chairs the Quarterly Leader Development Review (QLDR).[18] The QLDR is a collaborative decisionmaking forum, and its membership includes representatives from USAR, ARNG, Army Commands and ASCC, and HQDA staff principals. QLDR members critically examine leader development initiatives and programs, discuss issues, and draw upon their experience and judgment to advise the TRADOC commander.[19]

TRADOC's **Institute of NCO Professional Development** (INCOPD) serves directly under the TRADOC commander, receives guidance from the TRADOC CSM, and provides direction and oversight of the NCOES across the Army. It also integrates all actions and activities related to NCO leader development into the Army leader development strategy and serves as the NCO subject matter experts for the Army leader development enterprise.[20]

The **United Stated Army Sergeants Major Academy** (USASMA) operates directly under the TRADOC commander, but receives guidance from INCOPD. USASMA "assesses, recommends, designs, develops, and executes programs for NCO development and education through a systematic, synchronized, integrated plan which provides the enlisted force with a comprehensive, single point-of-entry portal for engaging in both professional military development and accredited higher education."[21] USASMA develops ALC Common Core DL POI and courseware and conducts this phase of ALC.

The **Combined Arms Center** (CAC) also plays a key role in ALC strategic management. CAC "designs, integrates and implements leader development and the Army leader development program [ALDP]."[22] CAC has a major subordinate organization, CAC **Leader Development and Education (LD&E)**, which is the primary TRADOC staff organization that supports development of leader development strategies and execution of ALDP. In this regard, it also provides guidance as to ALC content and execution.

---

[16] See HQDA, *Army Leader Development Program Charter Memorandum*, December 2007.

[17] See TRADOC Regulation 10-5, *Organizations and Functions, U.S. Army Training and Doctrine Command*, December 2009.

[18] This forum was originally called the *Prepare the Army Forum* (PAF).

[19] HQDA, *Army Leader Development Program Charter Memorandum*, December 2007.

[20] HQDA, *Army Leader Development Program Charter Memorandum*, December 2007.

[21] TRADOC Regulation 10-5, *Organizations and Functions, U.S. Army Training and Doctrine Command*, December 2009.

[22] TRADOC Regulation 10-5.

NCOAs at **proponent schools and centers** conduct the MOS technical and tactical ALC phases using resident instruction, DL, or MTT. The MOS proponent school or center commander commands these academies. The proponent develops the MOS phase POIs and courseware based on higher-level guidance, as well as its own understanding of the training needs of soldiers in the MOS.[23] While courseware development is done by proponent institution training development staffs, the NCOA Commandant has a major influence on the design of the course, and its instructors and staff often are heavily involved in supporting the development process.

Management Is Decentralized, Making Strategic Synchronization Difficult

Although the process we described may seem fairly centralized, the reality is that the majority of ALC decisions are decentralized; that is, they are made by one or by a subset of organizations in the governance architecture. NCOES is only one of a large number of important training and leader development programs and receives strategic-level focus only when major issues arise, such as NCOES backlog; or when initiatives, such as MTT, are proposed and resources must be found to support them. While decentralized decisionmaking has advantages, it complicates the integration of ALC goals and resources for the best possible overall strategic ATLD benefit.

To consider one key example, decisions concerning ALC course content are key strategic decisions, but these decisions are, in reality, decentralized to the proponent and NCOAs.[24] Guidance comes from many directions to the proponent. While there is sometimes specific higher-level guidance to add specific subjects, such as suicide prevention, most of the guidance is general. Examples include the NCOES transformation guidance to teach both SL3 and SL4 tasks in ALC and guidance to include recent lessons learned. While there is much guidance on what to add or enhance, little guidance delineates what content can be reduced or removed. This, coupled with the guidance to limit course lengths to no more than eight weeks, means that decentralized decisions must be made on what is added, taken out, or changed.

Moreover, there appears to be limited follow-up in terms of determining the degree to which guidance is implemented, and in general the judgment of the proponent is respected in regard to making course content decisions. For example, when we examined ALC Common Core and several MOS POIs, we found limited or no inclusion of SL4 tasks, even though the NCOES transition guidance was to focus on both squad and platoon leader skills. Also, the current ALC Common Core Phase focuses on squad leader skills and contains few platoon sergeant–level tasks. The proponents considered the guidance, looked at training needs, and made what they thought were the best decisions.[25]

Another example of competing goals relates to the strategic issue of ALC's relationship to promotion. Promotion policies are a DA G-1 responsibility. As mentioned, ALC was previ-

---

[23] Many but not all TRADOC proponents are under CAC. For example many support MOS proponents are under the U.S. Army Combat Arms Support Command (CASC). However as the leadership proponent, CAC provides general leadership guidance to all MOS proponents.

[24] During our visits to NCOAs, one question we asked was which governance organization provided them guidance; each time, the answer was, "All of them."

[25] See BNCOC Common Core DL POI, August 2008. The recently developed ALC DL Common Core contains the same tasks and focus. Also see U.S. Army Ordnance School, *91B30 Wheeled Vehicle Mechanic Advanced Leaders Course (ALC) POI*, April 2009, which focuses almost exclusively on SL3 tasks.

ously a requirement for promotion to staff sergeant E6. Recently, ALC has instead been made a requirement for promotion to sergeant first class E7. This change seems reasonable given the important consideration that a soldier should not be penalized because deployments prevented timely attendance. However, the decision does not square with other Army goals. Specifically, because promotion is a strong motivation for attendance, the change limits the incentive for senior E5s and junior E6s to obtain the institutional training designed to support performance in SL3 positions. Thus, making ALC a requirement for E7 promotion works against the goal of reducing ALC backlog (defined as E6 and higher soldiers who have not completed ALC). Nor does it fully support unit ARFORGEN needs. ALC supports ARFORGEN processes by providing the right training at the right time—in this case, institutional training—to ensure that staff sergeants have the technical and tactical skills needed to support unit collective training in the Train-Ready phase.

A related strategic issue for ALC concerns aligning resources with training objectives. The addition of sergeant first class skills to ALC is a major addition to course goals, but with no increase in course time (in fact the emphasis has been on reduction), the overall effect is that the number of tasks and skills needed for each skill level being taught has been reduced. Nor was this decision supported by increased instructor grades (from E6 to E7).

There also are examples of decisions affecting ALC being decentralized and not necessarily in concert with achieving overall strategic goals. These include decisions on use of DL and training development support to achieve the goals of best possible support to ARFORGEN processes and reduction on soldier and family turbulence.

Finally, we observe that decentralized decisions make it difficult for operational force commanders to have a full opportunity to participate in strategic decisions. Too many decisions are being made at too many locations and times.

Information Support of Strategic Management Has a Number of Gaps

As a key part of our analysis of strategic management, we examined the decisions that needed to be made and the information that would logically be needed to support informed decisions. We identified four categories of information: unit constraints, NCO improvement needs, costs, and benefits. We then compared the information actually available in these categories to that needed and drew conclusions concerning the adequacy of information availability. The results of this comparison are shown in Table 3.1. Our overall conclusion is that a large amount of the information needed for informed strategic ALC decisionmaking either is not available or not readily available to decisionmakers. Given limited access to needed information, decisions appear to be heavily based on the individual experience of senior leaders, anecdotal information, and assumptions.

Unit Constraints. A more complete understanding of the demands on, and needs of, units during the Reset and other phases of the ARFORGEN cycle would enable Army leadership to make balanced decisions about ALC lengths, modalities (e.g., MTT and DL), and reasonable options for reducing backlogs. There is a general understanding of the major activities needed for an effective Reset and their difficulty, but not of the specifics such as the time, difficulties, and levels of effort required for each, especially in terms of the need for unit NCO leadership. These and similar data could support more informed strategic decisions about what is reasonable in balancing ALC content, length, and modalities against unit needs for effective movement through the ARFORGEN cycle.

Table 3.1
Information Available to Support Strategic Management for ALC

Category/Item	Available	Partially Available	Not Available
Unit Constraints			
Available Time		X	
NCO Improvement Needs			
Individual			X
Unit Readiness			X
ALC Costs			
Dollars		X	
Course Time—Military	X		
Manpower Unit/Soldier		X	
Impact		X	
ALC Benefits			
Production of Graduates	X		
Percentage of Requirements		X	
Learning			X
Unit Readiness			X

NCO Improvement Needs: Individual and Unit. We found no systematic effort to collect information about which NCO MOS and general leadership tasks remain strengths or could benefit from increased inclusion or emphasis in ALC. A wide range of leadership surveys are conducted each year, but they provide limited value in identifying specific needs and directions for changing ALC.[26] Important areas include understanding, on one hand, the degree to which critical tasks and skills not being exercised during deployments are atrophying, and, on the other hand, how much that deployments are enhancing some traditional leadership skills and adaptability. There seems to be agreement that deployments are changing the nature of the contribution of operational experience to NCO skill sets, but we have found no structured examination of what this means in terms of how and in what areas ALC could best contribute to overall NCO proficiency.

Similarly, we could find no systematic, structured examination of unit operational performance to provide a basis for determining how ALC could be reshaped to support unit readiness improvement. Some guidance is provided based on understood force needs, such as stress reduction and emphasis on protection against improvised explosive devices (IEDs), but no systematic overall examination. Again, such information could provide decisionmakers a more objective basis for making decisions on direction of change for ALC content and emphasis.

ALC Costs. Many costs are associated with ALC, and some are difficult to identify. One that is directly visible is student time, because this directly relates to course length. The direct dollar and manpower costs allocated to operating the NCOAs also are visible in theory, but in practice difficult to aggregate. Actual resources used are not visible, because both TRADOC

---

[26] Both the Army Research Institute and CAC's Center for Army Leadership periodically conduct extensive surveys among Army personnel. These surveys generally provide useful insights on perceptions of leader competency and identify general areas where leader skills are seen as needing improvement, but not the specifics needed for the considerations being discussed here.

and the proponents have considerable latitude in shifting resources during the execution year, and actual ALC expenditures are only visible with considerable effort.

Another difficulty in gaining a full visibility of ALC costs is that many of the resources needed for ALC, such as costs of developing POI and courseware, DL courseware and support, and equipment support can be estimated but are not directly visible. This is because the activities that provide these kinds of support also pertain to a broader range of training and leader development activities, and the portion allocated to ALC is not directly visible.

The lack of direct visibility and a formal connection between programs can lead to strategic disconnects in resource allocation. One example is that NCOES is funded separately from DL. NCOES training strategies say DL should play a major role as a solution to the need to provide more training, to shorten institutional training time and to conduct PME at home station. Yet not nearly enough resources are devoted to the development of DL to support the continued development of these strategies to achieve their objectives in a reasonable time frame. In fact, funding for DL for major Army training activities has decreased in recent years.[27]

Another example is the fact that military manpower instructors in NCOES are funded separately from the Operations and Maintenance (O&M) resources needed for NCOA operation. As explained in the previous chapter, ALC was fundamentally changed recently from the predecessor course through the addition of sergeant first class skills to course goals. Yet the instructor grades have not been increased to provide for instructors with sergeant first class experience to teach the elevated skill sets.

Even less visible at a strategic level are the costs of unit ARFORGEN implementation; specifically, understanding the effects on unit Reset programs that result from the absence of a significant proportion of key leaders on recovery and regeneration activities.[28] Likewise, there are no captured unit costs in supporting MTT execution.

A later section in this chapter examines the complexities of managing the resources needed to support ALC.

ALC Benefits. As with costs, identifying the benefits of ALC is difficult. Some benefits of ALC are directly observable, others less so, and some generally remain invisible. The number of ALC graduates, an output that acts as an overall proxy for benefit, is readily available from the Army Training Requirements and Resources System (ATRRS). The fill rate, i.e., the ratio of enrollments to allocated quotas, is also easily available. The fill rate gives some information about the performance of the ALC management system and the likely increase in the backlog. However, it does not enable decisionmakers to compare the number of ALC graduates with the total number of soldiers requiring ALC, that is, pure requirements.

Moreover, while the number of graduates can be counted, determining exactly what training they received is harder to do. The length of the course gives some indication, but not in terms of tasks trained and familiarized. There are listings of tasks and learning objectives of each course available from POIs, but this information is not always complete or accurate, because proponents often change course content faster than the changes can be documented in POIs.

---

[27]   See Shanley et al., *Making Improvements in the Army's Distributed Learning Program*, March 2012.

[28]   For example, the typical time in grade between promotion to sergeant E5 and to staff sergeant E6 is around four years. In an AC ARFORGEN cycle of three years, half or more of the senior sergeant E5s, and junior staff sergeants could be in ALC and not supporting unit reset activities for up to eight weeks of the six-month Reset phase.

Student learning is even more difficult to measure. The inclusion of material on a given topic in the course does not mean that the graduates can actually perform the technical and tactical tasks in an operational environment, or apply the leadership skills with soldiers in a unit setting. Thus, the inclusion of a task or skill in a course, or even the fact that a student has passed a test, does not necessarily mean that better graduate performance will result. As the pressures mount for adding leader tasks and skills and for shortening course length, the need to understand actual learning benefit becomes increasingly important. But we could find no consistent method of measuring learning across proponent ALC courses, nor was there a mechanism to provide Army-level decisionmakers with an objective understanding of ALC learning levels.[29]

ALC benefit to unit readiness is another area in which strategic decisionmaking could benefit from improved information. Although ALC has long-term leader development goals, the wider range of requirements for both full-spectrum and deployment-specific training readiness have placed increasing importance on the ability of ALC graduates to contribute to implementation of unit readiness programs and to perform the technical and tactical requirements of unit duty positions. This need is accentuated by the constrained time unit commanders have to meet these requirements during the ARFORGEN Reset and Train-Ready phases. Thus, an important consideration is the level at which ALC supports near-term unit readiness and readiness preparation needs. However, given the decentralized nature of decisions on course content, objective benefit information of this type is not systemically available to decisionmakers above the proponent level.

Lack of Data Precludes Effective Cost-Benefit Analysis

Effective business practices require a cost-benefit analysis in some form. For ALC, only limited information is available on the benefits side, and there are major gaps on the cost side as well.

The cost and benefit tradeoffs involved in promising new training methods, like MTT and the innovative use of DL, cannot be easily explored with the existing Army analytical capability. Blended learning is one such approach, but there are no cost factors to support resource decisions connected with its implementation, and aggressive piloting to determine those factors has not been given strategic priority. Moreover, in an environment of scarce resources, there is a natural tendency to underestimate what new approaches might cost, and the lack of funding for implementation is likely to discourage new ideas.

Even with full visibility of existing costs, it is difficult for the Army to conduct "what if" analyses concerning strategic decisions on new directions for ALC. For example, we found no tool to look at tradeoffs between ALC costs and unit costs for conducting training, or costs of adding or subtracting to ALC at the task level.[30]

IT Systems Provide Limited Support to Strategic Management

Five main information technology (IT) systems support strategic management. As described above, these data systems can provide some, but by no means all, of the technology required to support strategic decisionmaking. The systems are ATRRS, the Army Personnel Data-

---

[29] AUTOGEN (Automated Survey Generator), which includes surveys of graduates and graduate supervisors, is a current attempt to provide some of what is needed, but it has low return rates, among other problems, and it has not yet provided useful overall conclusions.

[30] It should be noted that recent strides have been made in the development of analytical tools that could support strategic decisionmaking. These tools are described in the next section dealing with program management.

base (TAPDB), the Digital Training Management System (DTMS), the Individual Training Resource Model (ITRM), and the Automated Systems Approach to Training (ASAT).

ATRRS provides information about ALC requirements, graduates, and course length; it is the Army's system of record with regard to ALC and other course completion. It contains a wealth of data and could provide more if directed and funded to do so, such as a measure of graduations against the requirement.

TAPDB is the personnel management system of record, and contains data on soldier grade, current and previous assignments, and education.

DTMS is the Army's system of record supporting unit training management functions. It is a Web-based, commercial off-the-shelf software application customized to provide the ability to plan, resource, and aid in managing unit and individual training at the unit level. It has the capacity to be a repository for data detailing the nature of current unit training programs. However, many units do not currently use this system, and it would take great effort to acquire data from all units.[31]

ITRM provides a capability to estimate ALC costs in both aggregate and detailed terms for traditional residential ALC. However, given the lack of information about costs and benefits, it cannot fully support many strategic analyses. Further, it is somewhat limited by its sometimes outdated source material, and it cannot estimate the cost of new training approaches, such as DL and MTT. It does, however, have a budding "what if" capability with regard to the examination of new training scenarios (described later in this chapter).

ASAT provides detail on the tasks and skills taught in ALC (in POIs) and resources required for ALC courses (in Course Administrative Data [CAD]). It also contains general information on the total number and type of tasks in the current course. However, there are gaps in ASAT data and some data are incomplete or not current. Moreover, it would take more effort to obtain these data from the system in a form that would support strategic decisionmaking.

In general, the completeness and accuracy of the information that is supposed to be contained in these IT databases varies greatly and depends on funding and other incentives for maintaining and updating data fields. For example, POIs in ASAT are not necessarily current, nor do they capture actual tasks taught. DTMS also has much missing data. ATRRS is reliable as a system of record for training and education information, but not necessarily accurate within secondary data fields. For example, if analysts wanted to use student email addresses to conduct a student survey about the usefulness of ALC training, they would encounter missing and incorrect data for a noteworthy portion of the students.

Integration across data sets can also be problematic. In a few cases, pairwise connections have been established. For example, ITRM automatically pulls in data from ASAT. However, there is no connection between ATRRS and DTMS or ATRRS and TAPDB. This means that analysts often find themselves pulling data from several databases, addressing security concerns (especially when contractors are doing the work), verifying or "cleaning" the data, and making assumptions or estimates about missing but needed data—a labor-intensive effort. Even something seemingly as simple as calculating ALC backlog can require a fairly large effort.

Other Army data systems have some, but not all, of the data that could support ALC and other strategic decisionmaking. Moreover, some of the data are not directly useful for ALC

---

[31]   While DTMS is important to the support of strategic management decisions relating to ALC, it currently has no role in program management of institutional training.

strategic decisionmaking. Most importantly, there are limited data on unit training readiness and specific areas for improvement. The Army-wide system for this is the Unit Status Report (USR) in the Defense Reporting System-Army.[32] Commanders assess their unit's training readiness when completing USRs and report important resource constraints impacting readiness. But USR ratings are not intended to provide the specificity needed to support strategic, POI, and courseware revision ALC decisions.

Finally, some important data are not in any Army-wide database or available through an information system.

Strategic Management Activity Conclusions

Based on this examination of the strategic management activity, we drew the following three conclusions:

The complexity of ATLD strategic management makes effective synchronization and decisionmaking difficult. Strategic management must look across programs and allocate roles and resources and manage risk in a way that synchronizes their application and thus best achieves overall ATLD goals. For ALC, this means synchronization among ALC training goals, length, and methods; promotion policies; and unit and soldier recovery time needs in the ARFORGEN cycle. Balancing these somewhat competing goals requires difficult decisions to be made from an enterprise perspective. Decentralization, and the many organizations, forums, and activities involved in strategic management, make such synchronization and decisionmaking difficult. This complexity also makes it difficult for unit-owning commands to effectively participate in strategic decisionmaking processes.

The complexity of ALC strategic management also makes it difficult to make changes. Further ALC change is likely needed to meet changed strategic goals, such as increased support of unit readiness during ARFORGEN cycles, greater execution at HS, and widened learning goals. Only so much can be done within current resource allocations and available unit and individual time. This means that difficult, strategic-level decisions will be required to change in a way that achieves a more optimum balance of benefits and best use of resources.

Much of the information needed to make informed, objective strategic decisions is not available. An important example is the lack of information to allow strategic decisionmakers to have an accurate, objective, and current understanding of the training and leader development areas where improvement is needed. When coupled with the complex nature of the integrated relationship of the many ATLD activities combining to result in trained units and competent leaders, this further complicates strategic management of ALC. The lack of information and an analysis capability make it difficult for strategic decisionmakers to see the big picture needed to make difficult ALC decisions. Lack of information also makes it difficult to apply a cost-benefit approach to strategic decisionmaking. This is because little ALC benefit information is available, and there are also important gaps on the cost side.

All the considerations just described lead to an overall conclusion that strategic decisionmaking can only be improved to the degree that (1) an overarching strategic management architecture can be put in place; (2) the right information is available to support it; and (3) chain of command customer needs can be more directly considered. These conclusions are easy to state, but actual improvement will take time and effort.

---

[32] See AR 220-1, *Army Unit Status Reporting and Force Registration-Consolidated Policies*, dated April 2010.

POI and Courseware Development

The "As-Is" POI and courseware development activity for ALC is displayed in Figure 3.5.

Inputs and Outputs. The outputs are changed POIs and courseware. ALC POI and courseware are revised when guidance, training requirements, and resources change. The rate of these changes in today's operational and training environment is high, and generates a need for constant and often major revision. Strategic guidance, such as the decision to include SL4 skills in ALC, can generate the need for change. POI and courseware revisions are made based on proponent and academy internal operational analysis and assessments of areas where ALC improvement or adjustment may be needed.

POI and Courseware Development Process

The Army processes for POI and courseware development and revision are defined in detail in TRADOC Regulation 350-70, *The Systems Approach to Training Management, Processes, and Products*, and a series of supporting TRADOC Pamphlets.[33] The processes outlined are logical and systematic, but require a major, deliberate staff effort. The challenge in implementing the processes is that TRADOCs training development staffs have been resourced at far less than required levels for many years, while the current level of effort required to keep POIs and courseware current and relevant has grown exponentially.

TRADOC has recognized the need for improvement, and a revised process, called Army Training and Education Development (ATED), is being developed and has been partially implemented. The objectives of the revision are to emphasize development of enhanced education modalities, especially in regard to complex thinking skills, and to make training develop-

**Figure 3.5
ALC POI and Courseware Development Activity**

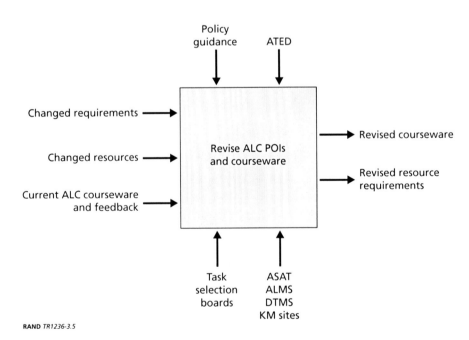

RAND *TR1236-3.5*

---

[33] This regulation is being revised to align with the new Army Training and Education Development (ATED) process described in AR 350-1.

ment processes more efficient. TRADOC institutions currently are implementing aspects of the new process, specifically in terms of emphasizing new educational models for developing complex thinking skills.

The ATED process has five steps, which in terms of ALC POI and courseware revision can be summarized as follows:[34]

- **Evaluation**: Determine how well the ALC training is done and how operational performance could benefit from new or improved NCO skills and knowledge.
- **Analysis**: Determine what NCO tasks and supporting skills and knowledge are critical and develop specifications and performance standards for these.
- **Design**: Determine whether the tasks will be taught or trained in ALC, and, if so, how this will be done.
- **Development:** Make any needed revisions to the POIs and courseware.
- **Implementation**: Execute the training using revised ALC POIs and courseware.

As outlined previously, a key ALC decision is what tasks and learning objectives should be included. There also is a need to determine whether revision of the course is necessary; what revisions are needed and feasible with the resources expected to be available; and how to implement those changes, including the selection of media to be used. Possible media choices are resident, MTT, and the various forms of DL.

Proponents also do cost-benefit analysis regarding course revisions. They typically consider the effort and resources that will be needed. They also consider the funds, instructor talent, and other resources they have received in the past and what they expect to receive in the future.

Faced with a need to make major POI and courseware revisions but with less-than-adequate training development manpower resources to apply the SAT process, the training development staffs for the ALC courses that we examined all applied modified, more informal, less supported approaches to POI and courseware revision, and these varied across proponents.

Controls. There are two main controls (i.e., means to guide, direct, constrain, or provide context for POI and courseware development). The first is the ATED process described above. The other is the strategic guidance that can come from the many different directions described in the previous section on strategic management.

Mechanisms. A key mechanism is the proponent Task Selection Board, which, according to current guidance, meets triennially or as required. The Board is convened by the proponent to review and update the task lists in TRADOC's Individual Training Plans and to determine which tasks will be accomplished in ALC.[35] During our discussions with school staff, we found that the Task Selection Board process has been outpaced by the size and rate of change, and changes have been made outside its framework.

---

[34] This is the ATED process outlined in AR 350-1. Basically it is the same as the Systems Approach to Training (SAT) process, which has five almost identical steps. The SAT process starts with evaluation, as opposed to analysis, but in a practical sense this does not constitute any real difference. The process is iterative: ALC execution is evaluated and the evaluation can then generate analysis and subsequent steps.

[35] TRADOC Regulation 350-10, *Institutional Leader Training and Education*, August 2002.

Four IT systems directly support POI and courseware revisions: the ASAT, the Army Learning Management System (ALMS), the Digital Training Management System (DTMS), and Army Knowledge Management sites.

ASAT is a tool that automates and facilitates the ATED process, including POI and courseware development. It is also a repository for training products, including POIs and supporting products. All these functions support POI and courseware revision. Its usefulness is undercut by the complexity of the SAT process and by a lack of resources to follow its steps, and the completeness of the repository. Both the degree to which ASAT is used and the degree to which training products are included in repositories vary by proponent.

ALMS is an integrated set of tools designed to support the management and execution of institutional training, including testing, development of course material, and analysis. ALMS does not have uniform usage across proponents, as many think that some parts of the system are not user-friendly and do not meet their learning management needs.

DTMS also supports POI and courseware development. As discussed in the strategic management section, DTMS has been developed and fielded, and it has the capacity (or capability) to provide courseware developers with information that could provide a better understanding of the nature of unit programs and their constraints and needs. However, few units enter the full range of data needed to provide this understanding. Moreover, even if the use of DTMS increases, it appears that further revisions would be needed to enhance its capabilities to where courseware developers could easily collect, aggregate, and analyze the data to understand these areas.

Knowledge Management (KM) Sites. POI and courseware development also requires access to information for content development. As seen with training development, there are staffing problems. A shortage of doctrine development staff has resulted in shortfalls in TRADOC's ability to maintain a reasonably complete set of doctrine, tactics, techniques, and procedures (DTTP) to support today's full-spectrum leader learning needs.[36] Lacking a full set of primary source doctrinal-level publications, courseware developers must rely on personal experience and a search of a wide range of secondary course material. There are myriad KM sites, including the Center for Army Lessons Learned (CALL) and the Battle Command Knowledge System (BCKS), and FORSCOM's Warfighter Forums, which have information that could support content development. The problem is the amount of effort required to find, analyze, and distill the material needed to support instructional revision from among the vast amount of available information. While a lack of data is a problem in itself, too much information in myriad differently structured databases can also make effective POI and courseware development difficult.

POI and Courseware Development Processes Have Changed to Adapt to Resource Constraints

NCO academy staffs voiced concerns that they are being asked, and need, to teach more in their courses without being relieved of any requirements or being given additional resources. While there was a general belief that academies are adjusting to what is being asked while still maintaining their standards, it was also felt that some compromises in student learning could

---

[36] The TRADOC commander has recognized the importance of this issue and directed a major effort to reengineer the Army's doctrinal development process.

not be avoided if the academies were to do more training within the same course lengths and with the same sets of instructors.

Because of training developer shortfalls, proponents rely on instructors for a major portion of formal and informal courseware development. Because experienced NCOs fill most instructional positions, this approach has typically benefitted ALC, and courses have been successfully adapted and modified on a regular basis.

However, the demands for change are coming from multiple directions, including requests to add common tasks skills to the MOS phases. This has complicated effective POI and courseware revision, because of greater need to balance MOS-specific and common skills in the MOS phases.

Across the NCOAs, course revisions are typically made as quickly as possible to implement guidance and keep instruction relevant and beneficial. While there was ongoing revision, often the changes were not captured in POI revisions. Generally, it seems that updating POIs was not given priority unless there was a need for increased resources, in which case course changes needed to be documented on POIs and CADs.

Many revisions are being developed by the academy staff, especially the addition of recent "lessons learned" or revised techniques and procedures to the course material. Formal and informal feedback from students is a key input for these revisions, since students sometimes have more recent operational experience than instructors and training development staff personnel.

Further Changes Within POI and Courseware Development Could Improve ALC Benefit

While we found significant ongoing modification of ALC POI and courseware, our conclusion is that further changes are possible and could improve ALC benefit and efficiency. We point out two possible areas that could be considered: (1) greater use of DL and (2) increased synchronization of course learning objectives with ARFORGEN training strategies. We point these out not to say that the Army should necessarily implement these specific changes, but to illustrate that significant changes should be considered if there is a possiblity of significant ALC improvement.

Expanded Use of DL. As mentioned above, DL has been little used to increase the exportability of the MOS phases of AC ALC courses to HS. In this regard, the use of MTTs has been acknowledged as a general success, but the use of this method has been extended about as far as is reasonably possible. DL has the potential to supplement MTTs and thus increase the amount of ALC instruction that can be exported to HSs, especially for courses with annual student densities too small to make MTTs cost-effective.

While many field exercises and hands-on equipment training modules are inappropriate for DL, most classroom learning could be trained in DL. This is especially true if the use of collaborative asynchronous DL is considered, such as with instruction in which there is interaction between students and instructors by means of threaded discussions or some other vehicle. An example would be DL modalities in which the student studies a subject area, then turns in a written product such as a warning order or Operations Order (OPORD), or participates in threaded student-to-student and student-to-instructor discussions online, or calls in and has live dialogue with an instructor.[37]

---

[37] Our dialogue with Special Forces and Armor Captains Career Course DL instructors indicated that such methods allow for reasonable student-to-student and student-to-instructor interactions. See Shanley et al., *Making Improvements to*

The movement to online instruction of this type is a major area of growth in the civilian educational community, and many NCOs have participated in such instruction. Collaborative DL also has the advantage of requiring many fewer training development resources. Moreover, development is often technically basic enough that an expert is not required.

Until recently, the Army's DL program has focused on the needs of the Reserve Component (RC) and on stand-alone IMI DL produced by a contractor. While the effect to date has been relatively small, the program is moving in new directions (including the increased use of collaborative DL and in-house production) that could expand its influence.[38]

There also is the potential to use computer-based instruction for "hands-on" functions that are done on a computer, for example "entering data into the Standard Army Maintenance System—Enhanced (SAMS-E) data system," or even for tasks such as "adjustment of fires techniques," which could be done in an online simulation. Another advantage of moving in this direction is that stand-alone computer-based instruction could be made available for refresher training or for training soldiers who need the skill but not the ALC course.

Even though few ALC courses may be suitable for full delivery by DL, there seems to be a far greater potential for courseware to move in this direction than has been accomplished. The goal would not be to eliminate all face-to-face or hands-on portions, but to limit it to those modules that can truly benefit significantly from direct contact. To the degree to which resident or MTT phases could be reduced, attendance in an ARFORGEN environment would be facilitated, and potentially the backlog could be reduced. We examined several ALC POIs and found that many lessons appear suitable for DL, especially if collaborative asynchronous methods and online games and simulations are considered.[39]

For progress in this direction, proponent training developers and members of task selection boards must understand the potential of various DL methods and the many benefits of moving in this direction. Because DL capabilities are advancing quickly, understanding them is a challenge, and many training developers and Task Selection Board members have a limited understanding of DL's potential and benefits.

Synchronization with Unit ARFORGEN Training. Another area in which POI and courseware revision could provide greater benefit would be increased synchronization with unit training programs and capabilities. For example, several of the courses we observed had training on Force XXI Battle Command, Brigade and Below (FBCB2) (a battlefield command-and-control system) as a part of the ALC. However, at each post there is a Battle Command Training Center (BCTC), a facility with a training staff for training soldiers to use battle command systems, including FBCB2. There might be better use of the limited ALC hours than training on a command-and-control system that the soldier may not use, or in which the version in the NCO's unit is different. Also, a larger number of ALC students would likely have expertise using the FBCB2 in actual operations. To the degree these factors occur, it might make sense to incorporate this training into unit training programs, rather than making it a part of ALC.

---

the Army's Distributed Learning Program, March 2012. The Army War College has used this same approach for nonresident instruction for years.

[38] See Shanley et al., *Making Improvements to the Army's Distributed Learning Program*, March 2012.

[39] For example, almost two-thirds of the ALC for 13F (Field Artillery forward observers) has classroom and Practical Exercise/test material that seems suitable for DL instruction. When we visited this course we asked several instructors if this was possible, and they agreed it was. As another example, a large portion of the 19K (Tank crewman) ALC is conducted in a classroom emphasizing tactical concepts and planning skills.

In an ARFORGEN cycle, an NCO who goes to ALC during the unit Reset phase will next go into small-unit collective training, the next phase of the ARFORGEN cycle. Prioritizing important "train-the-trainer" and tactical skills in ALC could help set the conditions for successful unit collective training. The difficult decisions concerning which leader skills to teach in constrained ALC POIs should be made in the context of synchronizing with unit needs and with unit training and leader development responsibilities.

Suggesting specific changes to ALC courseware is beyond the scope of this project. However, we suggest—along the lines of the reasoning above—that the process for developing POI and courseware could be revised and better supported. This would allow more informed and structured decisions concerning use of constrained course time and other resources to better support long-term leader development and near-term unit readiness goals.

Improved Information Is Needed for POI and Courseware Development Processes
Table 3.2 displays the results of our examination of the availability of the information needed for effective POI and courseware revision. The overall conclusion is that much of the key information is difficult for the proponents to get.

The limitations on availability of information for POI and courseware revision are somewhat less severe than those described in the strategic management section, but are still noteworthy. The ratings for NCO improvement needs and NCOES benefits are somewhat better, with many of the items assessed as unavailable for strategic management being rated as partially available for POI and courseware revision. The latter ratings are partly due to our judgment that the NCO academy and proponent staffs have a more direct understanding of current NCO strengths and weaknesses, as well as unit needs and capabilities, than do the strategic management staffs at higher echelons. This is because many ALC instructors and staffs have recently returned from operational tours. They also have a better understanding of course ben-

Table 3.2
Information Available to Support POI and Courseware Development for ALC

Category/Item	Available	Partially Available	Not Available
NCO Strengths and Areas Needing Improvement			
Leadership Skills		X	
Technical/Tactical Tasks/Skills		X	
Unit Needs and Capabilities			
Collective Training Needs			X
Unit Training Capabilities and Program Shape		X	
ALC Course Outcomes			
Content		X	
Levels of Learning		X	
Benefit to Graduate		X	
Benefit to Unit			X
Modality/Learning Methods/Content			
DL Methods and Capability by Type			X
Complex Thinking Teaching Methods		X	
Skill Decay Factors		X	
Operational Methods, Equipment, Systems		X	

efit to students because of direct contact with students during the course. The instructor who interacts with students during practical exercises, in classroom discussions, and during formal and informal course After Action Reviews has a far better understanding of student learning and course benefit than can be gained by reviews of student records and test scores.

An understanding of enhanced teaching modalities and methods is another area in which improved information would benefit POI and courseware development. The relatively low levels of DL use across ALC MOS phases indicate the need to better understand DL methods and approaches for developing and updating DL courseware effectively and responsively. DL capabilities and methods are continually advancing. Moreover, determining which tasks and skills can best be done by which DL methods is a complex undertaking that sometimes leads to conflicting views. These factors limit the capability of task selection boards, and even proponent training development staffs, to make informed decisions regarding the appropriate use of DL. Moreover, there is limited TRADOC guidance in this area.[40]

Similarly, understanding how to teach complex thinking skills is an area of growing importance. While there are many concepts as to how this can best be accomplished, there is general agreement that high levels of POI and courseware development and instructional skills are required.

A final area in which better information could support improved POI and courseware development is the determination of course content. Today's leaders require competence in a greater range of tasks and skills. Operational methods are changing to meet changing requirements. New operational systems are continually being fielded directly to operational forces. These factors limit the currency of proponent training development staffs and even its instructors. This rate of operational change has also surpassed TRADOC's capability to develop the doctrinal material that should be the primary source for course content.[41] This not only complicates the process of prioritizing course content, but also increases the workload to develop instructional support materials.

The issue of limited information is compounded by a corresponding limitation in staff training and the size of staffs involved in doctrinal development. Throughout our visits on this study and others related to training and doctrine development, the issue of inadequate staffing was a persistent and strong theme. While it is common for organizational staff members to think that they need increased resources, the combination of increased scope and complexity of course content, delivery methods, and required processes strongly suggests that this is an area of legitimate concern.

POI and Courseware Development Activity Conclusions

Based on this examination of the POI and Courseware Development activity we drew the following conclusions:

Further ALC POI and courseware changes could add benefit. ALC proponents are making major efforts to revise ALC POIs and courseware to make them relevant and beneficial to support unit readiness and longer-term leader development goals. However, further changes

---

[40] See Shanley et al., *Making Improvements to the Army' Distributed Learning Program*, March 2012. We examined TRADOC Regulation 350-70, *Systems Approach to Training*, and supporting pamphlets and found limited specific guidance to support selection of DL methodologies for specific tasks or skills.

[41] The TRADOC commander acknowledges this. See footnote 36 in this chapter.

to increase attendance at soldier HSs and for better synchronization with other ATLD strategies seem possible.

The ability to revise POIs and courseware is limited by a lack of information. Information about unit and force needs is needed to make the difficult decisions about what to include in the courses, but is not readily available. The information needed to make the training materials current and beneficial also is an issue, because operational methods are changing rapidly as are operational equipment and systems. The limited number of staff allocated to training development functions makes the lack of information even more of an issue.

Improved strategic decisionmaking is needed to support responsive POI and courseware change. While large ALC changes seem appropriate to better support the overall ATLD goals, ALC proponents have limited ability to make these internally. For example, schools can only go so far to implement DL without broader Army-wide changes and shifts in resources, such as increased DL development capabilities both within proponents and at unit locations.

Program Management

Program management activities focus on implementing and supporting ALC programs in a cost-effective way. Program management activities translate ALC strategic directions (e.g., conducting as much of ALC at HS during Reset as possible) and proponent recommendations for course revision into specific programs that are resourced to execute courses.

Subcomponents of this function are documented in AR 350-10, *Management of Army Individual Training Requirements and Resources*.[42] They include

- calculation of ALC unconstrained student requirements
- determination of the number of students the training system can support
- resourcing of the student training requirement in Army budgets
- scheduling of classes
- allocation of training seats to commands
- management of student reservations to particular classes
- storage of individuals' training history over time
- management of near-term training changes.

Program management involves a large number of decisions. DA's role in ATLD program management is to validate requirements, defend training programs, and provide and monitor resources. Examples of decisions DA has to support include the following:

- validation of critical student training requirements
- ongoing development and defense of budgets to provide resources for the training, including proposed resourcing of new training concepts
- validation and near-term adjustments in resources based on changes in requirements and changes in the resources available for training
- specific program direction to achieve strategic goals.

---

[42] AR 350-10, September 2009.

Each major sub-activity of program management is depicted in Figure 3.6, an IDEF0 process chart.[43] In general, the chart can be read as if it were two columns of information, from top to bottom on the left, then from top to bottom on the right. Each of the five "boxes" (i.e., sub-activities) of program management is described below.

As this figure indicates, program management is a complex, systematic, and comprehensive process involving many players. It has evolved over a number of years. Below, we describe each of the sub-activities in more detail.

Develop Unconstrained ALC Student Requirements

The output of this activity is unconstrained ALC student training requirements by MOS. About two years before the beginning of the execution year, each component determines the number of students that will need ALC, based primarily on the number of E6 positions in the force structure and the number of expected promotions. The effort is also informed by policy and program guidance related to those factors. Army G-3/5/7 Force Management Directorate (DAMO-FM) produces the force structure information, which provides the basis for the authorization documents used to determine ALC requirements.

These calculations are completed with the help of three models that take into account these and other inputs, with one maintained by each of the three components: Automated

Figure 3.6
ALC Program Management Activities

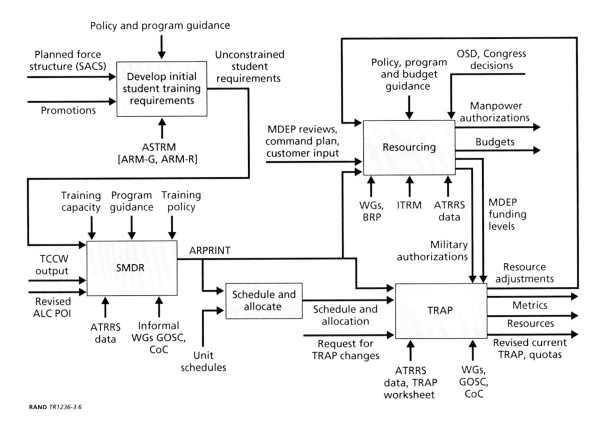

RAND *TR1236-3.6*

---

[43] See the introduction to this section for a description of how such charts are constructed.

Strength Requirement Model (ASTRM) is the model used by the AC, Automated Requirements Model—Guard (ARM-G) is used by ARNG, and Automated Requirements Model—Reserves (ARM-R) is used by USAR.

The requirements process depicted here could, in theory, also deal with the ALC backlog of student training requirements, although there is currently no defined process for doing so. In recent years, the backlog has not been considered in this phase of the program management process, because total student training completed has been below the level of unconstrained requirements.

Structured Manning Decision Review

The next activity (bottom left of the chart) in the Program Management Process is the Structured Manning Decision Review (SMDR), a process that validates the total ALC training requirement and reconciles that requirement with the capabilities of Army training commands, program guidance, and training policy. The ultimate objective is to determine an acceptable, affordable, and executable institutional training program for the Army.

Inputs and Outputs. The output of the SMDR is the approved Army Program for Institutional Training (ARPRINT) that will begin two years later. It is a constrained set of student requirements that schools agree that they can accomplish. For ALC, differences in the unconstrained requirement and the ARPRINT are relatively small; in 2008, the number of training seats in the ARPRINT was less than 5 percent smaller than unconstrained requirements.

The SMDR also forecasts a training program for later years (the POM years), and identifies critical adjustments needed for the training program that begins in one year.

The primary input to the SMDR is unconstrained ALC requirements. Approved ALC POIs are also inputs, reflecting school requirements on the length of the courses, as well as on training aids, equipment, and facility requirements.

Controls. In addition, schools prepare capacity reports that include descriptions of potential resourcing obstacles in meeting all requirements. Equipment and facilities are common resource constraints. Manpower is also a potential constraint, but has not been dealt with during the SMDR in recent years, since manpower decisions are made after the SMDR. Guidance is provided to the SMDR in the form of program guidance and training policy from senior leadership.

Mechanisms. The G-3/5/7 Directorate of Training and the G-1 Directorate of Military Personnel Management co-chair the SMDR. For purposes of ALC, the SDMR includes representatives from the Army staff, Army commands, National Guard and Army Reserve. These stakeholders work to provide solutions to requirements that schools initially determine they cannot meet.

During the action-officer segment, validated ALC student requirements for each course are compared with available training resources. Any course that lacks sufficient resources is termed "constrained." Solutions for such courses are sought using various strategies. For example, in some cases, resources are found after discussion with providers. In other cases, it is determined that commands have not historically been able to send enough students to meet the total requirement even when seats were available and adjustments were made. In still other cases, consideration is given to the option of "taking risk" with regard to certain training.

ATRRS data are an important supporting element for the SMDR because they show what training has been accomplished in the past, and therefore suggest what can be accomplished in the future. The data thus help align training resources with training need. The outcome for each constrained course is that either additional resources are provided or the training seats for that course are adjusted downward.

The action-officer forum is not a decisionmaking body. Thus, this segment of the SMDR is followed by an SMDR Council of Colonels (CoC) and a final General Officer Steering Committee (GOSC) that resolves the relatively few issues that cannot be solved at lower levels. The GOSC also approves the final training program as a whole.

Scheduling Courses and Quota Management

The ARPRINT is the official document that assigns missions to the training base and is the primary input for management of ALC student inputs or quotas. The outputs of this activity are a schedule of ALC courses by school and as assignment of course quotas to component commands that is then documented in ATRRS.[44]

This is a complex activity involving formal and informal collaboration among TRADOC schools, component commands, and HRC. For the schools, management of student inputs involves development of training schedules for all ALC classes. For components, management of training inputs means making use of the quota system, a process that allocates to various commands the right to fill a given number of training seats over a certain time.

ATRRS is the Army's management information system for managing student input to training. It is used to track individuals through the training base and to facilitate the filling of training seats. It stores class schedule information, provides the basis for quota management, accepts reservations by name and Social Security number for training seats, and stores enrollment and completion information for all students.[45] ATRRS also supports an evaluation of training program execution by capturing data and issuing reports on the fill rates that the ALC was able to achieve. In recent years, fill rates for the MOS-specific components of ALC have been relatively low—often less than 75 percent, depending on MOS.

Training Resource Arbitration Panels

Training Resource Arbitration Panels (TRAPs) (shown in the box to the bottom right of Figure 3.6) address unprogrammed changes to the training program after the ARPRINT is published. Most TRAPs occur in the year training is executed or the year leading up to the year of execution.

Inputs and Outputs. TRAPs involve decisions that lead to near-term changes to the ARPRINT's approved training requirement; the allocation of quotas that go along with those changes; (potential) training schedules; and the provision of additional resources for the additional quotas. TRAPs begin with units requesting additional training seats. "Offline" changes occur when changes can be made without involving the allocation of new resources. "Offline" TRAPs occur when there is a request for additional training resources, such as manpower, base operations support, equipment, and funding.

For ALC, TRAPs mostly deal with supporting MTTs at unit sites. TRAPs also occur as a result of force structure changes and mobilization training needs for the RC.

---

[44] Another output, not addressed in this report, is the scheduling of students to specific courses.

[45] ATRRS also supports special processes to aid training management. One example of its many modules that are important to ALC is the Unit Automated Reservation System (UARS), which is a scheduling aid for brigade-level commanders. This automated tool schedules individual training for soldiers in the Reset phase following redeployment. The BNCOC Automated Reservation System (BARS) is a complementary tool that supports scheduling of soldiers in garrison for ALC.

Controls. The budgets and the military manpower authorizations that resulted from resourcing the ARPRINT are the constraints considered during the TRAP activity.

Mechanisms. In cases where additional resources are needed, the G-1 (which manages TRAP changes overall) coordinates meetings with a TRAP action group to determine exactly what might be required. The TRAP action groups have representatives from the requesting command, the training component that would have to provide the training, and various DA staff representatives responsible for validating requirements or providing resources, such as those from DA G-1, G-3, G-8 and IMCOM. If a TRAP action group cannot resolve a TRAP issue, the G-1 presents the issue to the monthly TRAP CoC for resolution, and, if necessary the issue goes to the TRAP GOSC. When TRAPs are approved, training seats are increased or redistributed, and (where necessary) additional resources are approved. Other times, a decision is made to reject the request and "take risk."

Once a final decision about a TRAP has been made, a budgeting process starts to provide the needed resources. Depending on other budgeting priorities, obtaining the promised resources is not always assured, and, even when approved, receiving the resources often can be delayed by months. In fact, schools often have to juggle other resources to implement an approved TRAP at the time the training takes place, receiving the resources for that activity well after training execution.

In recent years, the number of TRAPs (and the number of associated training seats they represent) has increased dramatically, implying that more and more program management takes place in the near term, just prior to training execution. This increase has strained the TRAP system, which was set up to manage a relatively small number of "exceptions." It is not efficient to manage training by considering training requirements one by one and beginning the execution process before resources are confirmed. Such a process also can lead to a large number of false starts. For example, a fair number of TRAPs that dealt with setting up MTTs had to be rejected because resources (e.g., instructors) to provide the training at unit sites could not be made available in the short time available before the training had to occur.

As with other aspects of program management, ATRRS provides the starting point for training quotas and schedules. Calculating resources involved with a TRAP starts with an estimate from ITRM about what resources will be needed. However, because ITRM represents average costing, there is a need to pass around a "TRAP worksheet" to get exceptions and other differences from the average case from the schools involved. G-3 is then charged with bringing other data sources to bear to validate the TRAP request.

Resourcing

The resourcing box on the right side of the chart represents the multi-year activity for making resource decisions (e.g., manpower, budgets, equipment, and facilities) for executing ALC.

Outputs and Inputs. Final outputs are training budgets and military manpower authorizations.

The ARPRINT, describing validated training requirements, is a primary input used as a basis for developing ALC resource plans. DA uses the ARPRINT as one basis for building the POM. TRADOC uses the ARPRINT as one basis for developing the TRADOC Command

Plan, which includes identification of training resource requirements and services needed to execute the validated ALC training requirements developed during the SMDR.[46]

For Army programs, resources are provided through individually resourced subcomponents called Management Decision Evaluation Packages (MDEPs).[47] The primary MDEP that manages and oversees ALC is called TSGT (pronounced T-Sergeant for short, and not an acronym).[48] Basically, TSGT provides direct O&M funding for NCO academy operation, including civilian pay and operational tempo (OPTEMPO) dollars.

However, TSGT contains only a fraction of the resources required to implement ALC. The main resource needed to implement ALC, military manpower (instructors), comes from the Manning Program Evaluation Group (PEG). Another key resource that comes from outside TSGT is training development resources, which come from an MDEP directly devoted to that capability (Training Development [TADV]) and from another MDEP providing a distance learning capability (The Army Distance Learning Program [TADT]). Figure 3.7 shows the different types of resources and support ALC must receive in addition of those from TSGT

Figure 3.7
MDEPs and MDEP Groups Supporting ALC

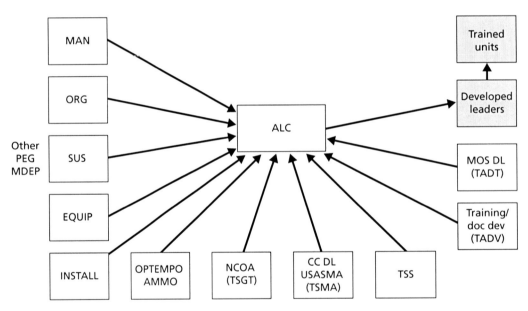

Supporting training PEG MDEPs (not inclusive)

RAND *TR1236-3.7*

---

[46] At the same time, other resource requests (besides those that flow out of the SMDR) are also fed into the resourcing process for consideration. For example, any unfunded requirement (UFR) at a school can result in the development of a concept plan that goes up through TRADOC Headquarters into G-3, which either validates or does not validate the requirement, and potentially builds the needed resources into the POM.

[47] MDEPs are specific line items in the Army's budget. Collectively, MDEPs account for all Army resources. See HQDA, *Planning, Programming, and Execution System*, AR 1-1, January 1994.

[48] TSGT supports other NCOES courses as well, such as SLC and Warrior Leader Course. Other MDEPs provide the primary support for training at other levels. For example, there is an MDEP for IMT, functional training, officer career development, and senior leader training. Still other MDEPs cut across types of training to provide specific capabilities, such as training development resources, DL, and various types of training support. In total, there are 122 MDEP managers in G-3 altogether. (HQDA, *POM 12-17, Update Brief to the November 2009 Training General Officer Steering Committee.*)

for the training to be executed. The figure includes MDEP and MDEP areas both within the Training Program Evaluation Group (TTPEG) (shown on the bottom right) and in other PEGs (shown in the column on the left).

Mechanisms. Beginning from the time the ARPRINT comes out, the process for all MDEPs starts with the building of the POM (completed by G-3) and then moves to a budgeting process (completed by the Army budget office). During the POM build, G-3 validates resource requests and determines which ones are critical. Of note is the fact that while NCOES traditionally has a relatively high priority within training programs, training overall is an area where the Army often decides to take risks.

As part of the process for providing resources for ALC, the G-3 looks at how many students historically have shown up for training so that they do not program resources for training that probably will not be needed. These processes take more than a year. Although budget numbers are constantly changing, there are two "lock" positions in the programming and budgeting process. One is called the POM lock, which is the resourcing position when Programs, Analysis, and Education (PA&E) turns the program over to Army Budget Office (ABO) for budgeting. The second, the POM/Budget Estimate Submission (POM/BES) lock, is the foundation for the budget that is sent to Congress for the next phase.

After the second lock, a proposed budget is sent to the Office of the Secretary of Defense (OSD) and then to Congress for specific decisions and guidance regarding funding. That process ends when Congress approves a budget, typically just preceding the beginning of the year of execution. As part of the process of following Congressional guidance, ABO takes congressional decisions, which are specified in terms of funding appropriations, and puts together specific school budgets and manpower authorizations for training organizations at Army Budget level of detail (e.g., the MDEP, Command and Army Program Element [APE] levels of detail). The time from POM/BES lock to the time that the budget is approved is nearly a year.

Budget recommendations developed at a variety of levels are made through a series of working group meetings. The MDEP manager for TSGT represents ALC's needs within these meetings. Working group meetings often include representatives from the PEGs (in addition to the MDEP manager), the rest of the Army staff, and the training commands.

Working group meetings feed into the Budget Requirements Process (BRP) group. The BRP is composed of representatives from G-8, G-3, and the ABO. They typically make integrating budget decisions. As with the SMDR, the resourcing (and BRP) process is also supported by higher groups that resolve disputes. These include a CoC, a two-star forum, and a three-star forum that can meet as often as once a week. Finally, there is a Senior Review Group (SRG).

ATRRS is an important tool in resourcing activities. It provides the history of training execution to inform decisions about how much to fund. In addition, ITRM provides a key building block for the POM, because it estimates how much a given training load would cost in budget and other resource terms. ITRM is a network of models that constitute the pricing mechanism for training program management. As a bottom-up approach to defining the cost of institutional training, it begins with force structure and workload, documenting the resources required for those needs. It then prices out the resources in a way that maps into Army budget categories (e.g., MDEP, APE, appropriation, command). Because it is built from the bottom up, it is also capable of looking at alternative scenarios in the decisionmaking process. In addition, the bottom-up approach provides the capability to link macro-level decisions to detailed impacts of those decisions.

Controls. Guidance during the budgeting process comes from many sources. For example, the program guidance comes from PA&E during the POM build. Budget guidance comes from the ABO. Broadly based policy guidance comes from many sources, e.g., DA staffs, the Chief of Staff, OSD, and Congress. PA&E summarizes policy guidance from many sources in the Technical Guidance Memorandum. While specific guidance concerning ALC would be rare, that training might well be affected by broader guidance affecting a larger portion on training (e.g., all of PME).

The ALC Resourcing System Is Fragmented

The first piece of evidence supporting this conclusion is that, as shown in Figure 3.7, the MDEP that manages ALC, TSGT, contains only a fraction of the resources required to implement that training.

For example, the most important resource for ALC, accounting for the largest proportion of total training cost, is military manpower (who serve as ALC instructors), and yet it is managed outside the TTPEG.

The manager of TSGT has limited control over, or even visibility of, many of the external resources needed to execute ALC. Military manpower can again provide a case in point. There are four different manpower levels relevant to the resourcing construct and ALC decisions: required, authorized, assigned, and actual manning. Manpower *requirements* are calculated using formulas that calculate instructor contact hours (ICH) required for each ALC module. While the total number of ALC ICH is contained in ITRM and in POIs by module, these figures often are out of date because they are no longer monitored or updated by TRADOC studies.

While *authorized* military manning figures supporting NCOES course execution are also recorded and available within ITRM, the decision process for authorizations does not typically consider ALC training issues, such as the higher instructor grades to support expanded ALC training goals. Moreover, the MDEP manager for TSGT has so little control over decisions related to authorized manpower that authorized levels are not even considered in the TSGT (or other training MDEP) briefings.

Military manning actually *assigned* to TRADOC is largely determined by HRC (as well as by G-1 and larger Army priorities) and varies by MOS depending on the demand for that MOS in operations. The distribution of assigned military manpower among NCOAs is managed by TRADOC and proponent commanders and generally not considered at DA level. Simply obtaining data on assignments would require a special request and a significant research effort.

Finally, exactly how much military manpower is actually dedicated to the support and conduct of ALC (*actual manning used*) is determined by individual TRADOC schools and is generally not recorded, nor even known, above the proponent school. Similarly, the unit effort to support MTT ALC is not recorded. As a result, a special and major research effort would be required to determine what it actually takes to execute both resident and exported ALC.

Fragmented Resourcing Processes Complicate Responsive Support to Implement Needed Change

It is important to emphasize that the limitations of the resourcing system cited above become particularly critical only when major change is contemplated and when the state of "change" is continuous. The PEG and MDEP systems work reasonably well in stable periods. For example, it is probably safe to assume that military manpower and dollars for training do not greatly affect each other when only marginal changes are contemplated. When coordination is needed, the relationships between MDEPs can evolve over time to a system of integration that works, just as the Military-Specific Training Allotment (TTDY, which funds student travel) has been coordinated with ALC and all residential forms of training. However when the change needed is large and rapid the informal system of coordination becomes inadequate.

Thus, if any of these resources funded outside of TSGT play an important part in transforming ALC, then that transformation is likely to be successful only if there is full coordination and integration among the various MDEPs involved. Yet we found that coordination and integration is difficult when implementing new methods of training, e.g., transferring travel funds to TRADOC to support MTT.

Even the basics of coordination can rapidly become complex when implementing training change. For example, suppose that as a precursor to designing change for ALC, program managers wanted to examine the existing distribution of manpower in NCO academies compared to the training load and determine how that distribution might be changed to increase efficiency and effectiveness. A full analysis would suggest an examination of three types of manpower—military, DA civilians, and contractors—because the different types can sometimes substitute for each other in the execution of specific missions within the schools.

Yet such coordination would be next to impossible. Civilian and contractor manning are funded out of the O&M appropriation inside TSGT, yet there is often limited visibility on the number of contractors, and changing civilian spaces would require actions and decisions outside of TSGT. Military manning, as described above, would be out of the DA training program manager's reach for purposes of coordination. Levels of support manpower used for ALC would require coordination with Training Center Operations (TATC), the MDEP that funds the larger TRADOC schools that supply that support. How much support manpower was used by NCOAs would be unknown without special school-by-school studies. Finally, the borrowed military manpower used by MTTs would not be visible without special studies.

To implement specific ALC initiatives, there are typically only informal and indirect mechanisms to coordinate and integrate funding outside the NCOES program. A key example, as described in the strategic management section, would be coordinating ALC with DL. While DL could be expected to play a major role in conducting ALC at home station, resources devoted to the MDEP for DL, TADT, are not integrally tied to future TSGT funding.

Even if TSGT integration could be achieved with the DL MDEP, efforts would rapidly extend beyond those two MDEPs to put together specific blended learning strategies. For example, consider a strategy for putting together an alternative to residential training that involved the following elements. First, structured self-development would get students to the "crawl-walk" level of training, likely using computer-based instruction and some level of instructor support. Then the training would move to more advanced simulations or gaming to get NCOs to the "walk-run" level of training, allowing them to "hit the ground running" in a shortened MTT training phase that would focus only on the most complex and hands-on aspects of tasks. To coordinate the implementation of this scenario, the TSGT would need to work closely with the training development and DL MDEPs even for the first phase. MDEPs that have to do with simulations and gaming would need to be brought in for the simulations piece. Coordination with the Manning PEG would be required to build in instructor support for these activities. Also, coordination with the U.S. Army Network Enterprise Technology Command (NETCOM) would be needed, because bandwidth (needed for the simulations) varies by installation.[49] Finally, arrangements and resourcing would also be needed to allow soldiers to take the DL and MTT portions of the training from home (e.g., for facilities and computers).

Much Important Information Needed to Support ALC Program Management Is Not Available or Is Difficult to Obtain

As with the other architectural levels, we examined the information that would be needed to support informed decisions for program management. We identified four categories of information: training management support information, costs, performance/benefit, and other information needed to inform ALC program management decisions. We then compared the information available in these categories to that needed and drew conclusions concerning the adequacy of information availability. The results of this comparison are shown in Table 3.3. Based on this comparison, our overall conclusion was that only some of the information needed to support ALC program management is readily available, and much is either not available or difficult to obtain. Note that two of the categories, costs and benefits, overlap the same categories used for strategic decisionmaking. While the strategic and program management areas

Table 3.3
Information Available to Support Program Management for ALC

Category/Item	Available	Partially Available	Not Available
Training Management Support			
Scheduling Data	X		
Quotas	X		
Reservations	X		
Course and Student Information	X		
Costs			
TSGT POM $ Amounts	X		
TRAP $ Amounts		X	
Manpower Costs		X	
Cross-MDEP Cost			X
New Design Cost			X
Performance/Benefit			
Production Fill Rate	X		
Production Compared to Resource Programmed		X	
Production Compared to Actual Costs			X
Resource Change Effects on Quality			X
Learning			X
Other Data			
Equipment and Facilities		X	
Deployment Schedule		X	
Force Structure		X	
POI Information		X	
Other		X	

---

[49]   NETCOM is the Army organization that plans, engineers, installs, and operates Army cyberspace.

often use the same databases, the former is more concerned with aggregate outcomes, while the latter is more concerned with outcomes in specific parts of the ALC system, such as at the command, school, and course level of detail.

Table 3.3 shows the availability of training management support, cost, performance/benefit, and other information. The categories of data are shown on the left, and the availability of data is shown in the three columns on the right. Below we further discuss information in each category.

Training Management Support. As described in the last section, we conclude that most ALC management support systems designed for training implementation work well from an enterprise point of view (see the "x's" in the "Available" column). Data stored and maintained in ATRRS that support training management include course information, class schedules, quotas, a reservation system, and course and student information.

Costs. Availability data for various data and analytical capabilities used in calculating training cost appear in the top section of Table 3.3. Much of the data to determine costs are readily available for the POM build, because ITRM supports the TSGT MDEP in the programming process for resident courses. However, ITRM is of less value in TRAP processes. While it does provide a starting point for many TRAP changes, its values do not necessarily apply in individual situations or when schools ask for special changes. For example, embedded cost factors of ITRM do not necessarily apply in TRAP situations. The marginal cost of training can be higher than the average costs in ITRM, and changes made at the last minute can be more expensive than changes that can be planned in advance.

Furthermore, the uniqueness of many TRAP situations can impose a difficult validation requirement on DA staff when it comes to determining the cost of training. For example, if a new training requirement has to be implemented by a contractor rather than by traditional methods, it may be difficult to determine whether the specialized tasks involved are appropriately priced by the school with the short decision cycle of a TRAP.

As described in the last section, a further problem is that ITRM neither reflects actual manpower costs used in ALC training nor the costs of support from many MDEPs that indirectly support ALC. Of even greater importance, the costs of institutional training to units, in terms of support that has to be provided (e.g., when MTTs are used) and leader time lost in unit programs, are not considered. Currently, capturing these ALC costs is simply not feasible in the fast-paced environment of DA decisionmaking.

Further, the costing of new paradigms like MTT and the innovative use of DL is not currently possible with ITRM. For example, using MTTs could reasonably be expected to result in a different set of costs than centralized training, yet no factors are built in that recognize the difference. Blended learning holds great promise, but factors for its implementation do not exist, and piloting to determine those factors has not been funded or proposed. In addition, in a resource-constrained area such as training, there is a natural tendency is to underestimate what such innovations might cost, yet underfunding is likely to provide a disincentive to those with new ideas.

Performance/Benefit. To defend their programs, argue for additional resources, or design improvement initiatives, ALC managers need to be able to monitor ALC benefits to students, chain of command customers, and the Army as a whole. Varying benefit levels also need to be weighed against the cost to produce them. Currently, the only metric easily available that relates to ALC benefits (or benefits in relation to costs) is the fill rate and graduation rates for ALC courses. Fill rate is the number of enrollments divided by total quotas, in which quotas are derived from validated requirements cited in the ARPRINT and updated in further processes. In recent years, the fill rate has been relatively low for ALC; in 2008 it was computed as 77 percent for MOS-specific ALC courses.

While the fill rate has value for program management (e.g., less than 100 percent indicates that backlogs are increasing and measures the extent to which full benefits are not being received for the resources expended), it is an input/output measure and falls far short of a complete measure of training system performance. For example, as noted in the strategic management section, there are no large-scale Army efforts to measure the learning of graduates or the connections of the course to unit readiness. Metrics such as these, while difficult to develop, would be more valuable outcome measures. Without some measure of what graduates have learned and how satisfied customers are with the outcomes, there is no way to relate training costs to the benefits achieved, e.g., to determine when continual resource cuts cross over from improving efficiency to decreasing learning.

Current factors do not reliably measure the performance of the schools that implement ALC. Information on school performance could provide a basis for introducing incentives for schools to increase that performance. Currently available data appear to support the contention that the school system performed reasonably well. For example, assigned manpower in the academies within TGST in 2008 was found by our analysis to be around 60 percent of manpower requirements, and MDEP briefs show TSGT to be funded at 50 percent of required funding for the ARPRINT requirement. These figures compare well with the 77-percent fill rate cited earlier, making it appear as if schools exceeded expectations.

However, judging school performance using currently existing data runs the risk of misinterpretation, because there is not a well-established relationship between resource levels and the quality of training and the number of tasks trained. While the schools seemingly conducted considerably more training than they received resources to execute, the current fragmentation of the training resource process makes other explanations equally plausible:

- Since actual costs are not monitored, the outcomes for ALC might reflect TRADOC commanders' decisions to commit more resources to ALC at the expense of other, unknown activities.
- More training might have been completed but at the expense of a lower quality level (e.g., fewer tasks trained, less student learning) or an unsustainable Personnel Tempo (PERSTEMPO) for school personnel.
- The apparent success suggests that the ICHs resourcing factors could be inflated.

Finally, in the current ATLD environment there are many unresourced requirements, and continual cuts have to be absorbed within a very short time. Except for some limited capability within ITRM (see further explanation below), analytical tools are not available to perform "what if" analyses when searching for solutions to particular problems involving outputs, benefits, and costs. For example, the effect of a decrease in spaces in TRADOC on the quantity and quality of training cannot currently be determined. As a more specific example, authorized military manpower changes cannot be tied to the potential effect reductions might have on training output.

The inability to quantify readiness impacts is a major issue for all ATLD programs. It makes it difficult for the TTPEG to compete with other PEGs, such as the equipping PEG, for which there is a much better connection between resource cuts and the consequences in terms of specific types of equipment that can be purchased or maintained.

It should be noted that recent strides have been made in the development of analytical tools for training within ITRM, and we cite these as efforts that need to be extended and expanded. For example, one model in the ITRM system is the Training Doctrine and Development (TD2) model. TD2 supports the TADV MDEP by connecting Training Development (TD) outputs with funding levels. It specifies the products that need to be used, the man-years of effort needed to produce them, and the priority for each class of document. While not a measure of customer satisfaction or benefits *per se*, the model can specify exactly what training development outputs would be sacrificed in the event of a funding cut. Eventually, the model might be incorporated in the manpower requirements process that produces school Tables of Distribution and Allowances.

Another ITRM model is the analytical workspace model (AWS-M) within ITRM. Available to HQDA, Major Army Commands, and schools, AWS-M is an evolving Web-based decision support capability that rapidly pulls in critical information in a form that enables and enhances analysis (e.g., by associating program funding with specific training outputs), including multiple "what if" scenarios. If appropriately funded, the model has the potential to develop into a large, generalized potential for cost-benefit analysis within institutional training.

Other Key Information Needs. Action officers often need rapid access to information both inside and outside the training system. Without visibility of the inter-relationships among such resources as equipment, OPTEMPO, and manpower, it is difficult to influence decisions. Above, we discussed those types of problems in relation to MDEPs. Here, we further discuss such problems in relation to particular types of resources and other levels of budget detail.

To cite one example, consider a request for additional ALC training that would require more end-items of equipment (e.g., tanks, helicopters). A DA training staff action officer might check with the DA G-4 staff equipment database and determine whether any items of equipment might be available for the training. If none are available, it may look like the training cannot be implemented. To verify this, the MDEP manager may need to know the number of end items that are currently available on particular installations to determine whether a solution might be feasible.

Information is thus needed not only about ALC training itself (e.g., POI information), but also about equipment and facilities, deployment schedules, and force structure (see Table 3.3). In addition, analysts may need various levels of detail, not just the level of detail available in the budget. For example, with all NCOES courses in one MDEP, it would be difficult to look at an issue involving one or only a few ALC courses.

Databases often seem to contain the desired information, but it is difficult to get data quickly without expert help. Data often are stored in multiple and overlapping systems, and each system can have different rules as to access and different formats and methods of extraction. Moreover, results are often not easy to interpret. As a result, this type of information often is not available to G-3 in a timely enough fashion to support specific decisions. Without a common view of requirements and resources (in near-real time), action officers often have to make decisions with far less than perfect information.

IT Systems Supporting Program Management. As described above, requirements models and the ATRRS and ITRM data systems are the main IT systems that support program management for ALC. Other systems also play a role, as feeders to those systems (e.g., ASAT feeds ITRM) or as external data that contain information that can influence training outcomes (e.g., on equipment, facilities, deployment, and force structure). These systems currently provide some, but not all, of the data that would be required to support program management during periods of transformational change. The greatest needs during these periods would be for better data on training benefit, on the cost of new training approaches and MDEP interaction, as well as more timely data inputs to support program management processes. In addition, greater integration is needed among training data systems and external data systems that deal with such factors as equipment, facilities, and deployments.

Program Management Activity Conclusions

Based on this examination of the Program Management activity, we draw the following conclusions:

Program management works well for traditional resident ALC. While the existing program management activity for ALC is complex and requires considerable effort and support for its operation, it works reasonably well for implementing a well-defined institutional ALC program. For example, defining unconstrained requirements for ALC is a relatively straightforward undertaking. The number is driven by well-defined promotion projections and well-defined models.[50] A large number of critical training management support decisions, supported by ATRRS, also work well. These include ALC scheduling management, quota management, and reservation management. In addition, ATRRS does a good job of recording throughput information about courses and students. The ITRM model also supports calculating the dollar amounts required for resident ALC budgets.[51]

The ALC Program Management system is not set up to support major change. While the system operates well in relatively static situations, it tends to be stressed during periods of change, especially when changes occur within shorter and less predictable timelines than they have in the past. For example, even the seemingly simple move to executing a portion of the ALC courses by MTT, as well as the synchronization of ALC schedules with unit ARFOR-GEN cycles required large, special efforts. To cite another example, the system is not set up to support larger changes, such as greater exploitation of DL to support anywhere/anytime training. In this regard, of great importance would be improvements in integration in the PEG and MDEP systems to deal with what can become a fragmented resourcing process when major changes are contemplated. In addition, a greater analytic capability is needed to support the assessment of costs and benefits of the proposed changes.

---

[50] Note that this process as it applies outside of ALC, such as for IMT and functional courses, is likely to have even greater problems.

[51] Some of these systems are more stressed when it comes to implementing ALC at home stations using DL and MTT as vehicles. These issues are further discussed below. Also, increased frequency of change has placed increasing pressure on some of these IT systems.

Execution

As displayed in Figure 3.8, there are two main execution activities—scheduling and getting students to ALC and execution of the ALC POI by the academies.

Student Scheduling and Attendance

HRC centrally manages scheduling AC soldier attendance at ALC. The activity would seem to be fairly simple and direct, as HRC knows the quotas from the ARPRINT as well as soldier ALC eligibility and unit of assignment from the TAPDB. It also knows the unit deployment schedules. However, the coexistence of a high backlog combined with relatively low fill rates for ALC in recent years is evidence that in the current era this activity is indeed difficult.

This difficulty occurs despite the many efforts of HRC to resolve scheduling issues. The HRC staff has to manually integrate three databases: TAPDB, ATRRS, and deployments schedules. Another key example of HRC's efforts has been making direct coordination with deployed units 180 days prior to their redeployment to verify the names of soldiers eligible and available for ALC. ALC MTTs have been another initiative to reduce scheduling issues.

Many factors appear to complicate effective scheduling. The most dramatic is that the windows for attendance have been narrow. Soldiers cannot be scheduled for activities away from HS until 90 days after their return from deployment, and units have often been scheduled for their next deployment only a year after their return. With such narrow windows for recovery, training, and other important preparation activities for the next deployment, many commanders are naturally reluctant to send a major portion of their junior NCOs to ALC. Also, many low-density MOS ALCs are scheduled only a few times a year, further complicating scheduling to balance ALC attendance and more direct unit-readiness needs in this narrow window.

Another issue is that many soldiers scheduled for ALC appear to become ineligible for the course. In some cases this stems from a physical problem, such as an injury occurring after

Figure 3.8
ALC Execution Activity

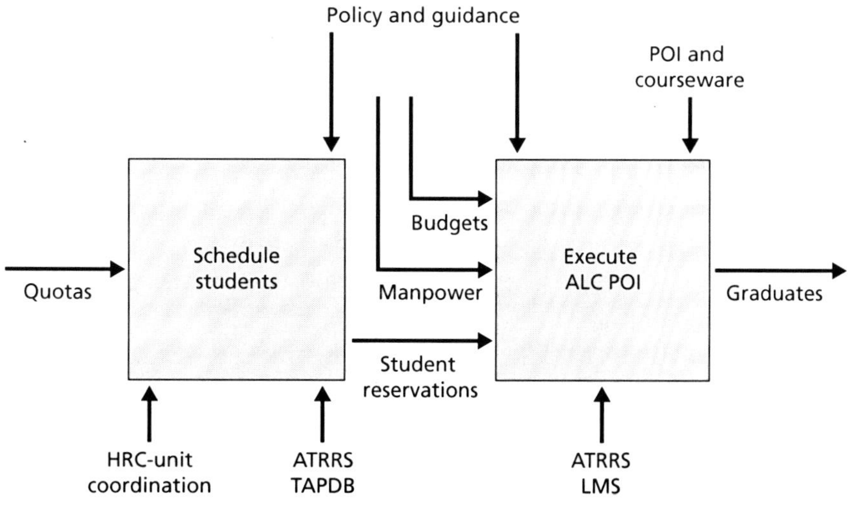

scheduling.[52] In other cases, the soldier decides to leave the Army. Also noted were cases in which the scheduling was a mistake—for example, the soldier had already attended ALC.

The disconnect between training and promotion policy is a contributing issue. Under the training policy an E6 is supposed to be an ALC graduate because ALC teaches critical E6 skills, but the promotion policy is that ALC is a requirement for promotion to E7—thus, given conflicting priorities, there is only a limited incentive to schedule a soldier to ALC before promotion to E6 from a promotion perspective.

While short unit dwell-time complications for scheduling soldiers to ALC are going down, the Army expects "persistent conflict" operational requirements for deployment to remain.

School Execution

Our research did not evaluate school execution of ALC. Instead, ALC execution was examined for the purpose of identifying how improved management processes above execution level could increase overall ALC benefit.

Our research indicates that ALC execution by staff and instructors is an overall strength. The professionalism and dedication of the NCOA staff and instructors are impressive, and survey data collected by the Center for Army Leadership show that a large majority of graduates see the course as useful and believe that it provides quality leadership development.[53]

One of its strengths is adaptability. All of the NCOAs reported that they were constantly adjusting the content of their lessons to keep them current with operational practices and needs. All had active programs to get student input to improve the courses, using surveys or formal or informal After Action Reviews, and often all three. They were able to make internal assessments and improve course material very quickly.

The degree to which the academies have adapted to support MTT also has been impressive. Our review indicated that the academies have implemented this method of delivering instruction to the extent feasible. They have developed effective procedures for coordinating execution with the chain-of-command customers they are supporting. Overall, ALC MTT processes appear to be working reasonably smoothly for such a major change. There are still issues with facilities and equipment, but these are also present for unit training programs, so it is likely they will not wholly disappear.

The adaptation is possible because execution is decentralized, which is a strength because it allows the academies to keep course material relevant and current. But adaptability also makes it difficult to understand course benefit and student learning at anything other than a general level for MOS ALC tasks and learning objectives.

Information Systems

Three main types of IT systems support execution: ATRRS, TAPDB, and school Learning Management Systems (LMS) including ALMS. All appear to be effective for execution management, but the capture of execution data has many limits for management decisions outside of the academy and ALC proponent.

TAPDB is the Army's personnel management system of record and provides an integrated human resources personnel database. Its contents include soldier contact information and data

---

[52] As an example, after redeployment a soldier scheduled for ALC may be suspected of having post-traumatic stress disorder, and treatment becomes a priority.

[53] See Center for Army Leadership Technical Report 2009-1, *2008 Leadership Assessment Survey Final Report*, May 2009.

related to grade, promotions, assignments, deployments, training, and education. Thus, it can easily be used to determine eligibility for ALC and aids in scheduling of students into courses.

ATRRS is the Army's system of record with regard to institutional training courses. Although ATRRS requires skilled operators to effectively manage student attendance, it is effective overall at scheduling students into the course, recording course completion data, and getting student completion into student personnel files. Some data fields that are not critical to the system's overall purpose (e.g., student email addresses) may lack accuracy. Other potentially useful data fields currently are not in use—specifically, the data on why students did not arrive at the course. The school is required to input these data, but often it has no direct understanding of the reason. This has hampered the Army's ability to diagnose and improve backlog and no-show rates.

LMSs vary across NCO academies, as does the information contained within them. For example, updated POIs and training support materials may or may not be in the proponent LMS. There is limited information concerning the specifics of student learning, that is, specific learning objectives, skills and tasks taught, and degree of student learning. An LMS may or may not contain all test scores, and if it does contain all, these may be kept on multiple systems, for example, scores for written exams on one and practical exercise–type tests on another. Moreover, there are likely significant differences in difficulty across tests and proponents. These and other inconsistencies limit the usefulness of LMS data and increase the difficulty of collecting data from the various LMSs.

ALC Execution Activity Conclusions

Based on this examination of the Execution activity, the following conclusions are drawn:

ALC is well executed. The courses have been adapted to support changing requirements to the degree possible given academy resources, capabilities, and guidance. However, the capability of the academies to make major changes within these parameters is limited. For example, shifting to a greater level of DL would require top-down guidance, resources, and orchestration. Likewise, gaining a full understanding of changing operational needs and methods and unit constraints can be done only to a limited degree at the academy level, given the resources available and the rates of change and operational complexity.

The level of success of the scheduling student activity has been less than desirable. This is evident by the backlog and unfilled quotas, but not because of lack of emphasis or effort on the part of HRC or the ALC academies. Better integration of IT systems could ease difficulty. However, scheduling students to relatively long TDY courses in an era of persistent conflict has no simple solution from an Army Enterprise perspective. It seems reasonable to state that there is a need to reexamine the priorities and balance among career development, unit readiness, and soldier care (promotion-criteria fairness)—needs that are above and outside of ALC Execution activities.

Decision and Management IT Support

In our case study, we analyzed the "As-Is" IT architecture that currently supports the ALC management and execution activities described in the earlier sections of this chapter. This IT architecture consists of the structure, components, and implementing technology of the collection of systems, services, and databases as well as the relationships and connections among these elements. It also includes the organizations and processes that are involved in developing and maintaining the elements of this architecture.

This IT "architecture" is not a formal architecture in the sense of being designed as an entity. Instead, it came into being in an ad hoc manner, generally as separate databases or services designed to accomplish distinct functions. What we examined were the de facto capabilities, arrangement, and relationships of IT systems that we found in support of ALC management.

To understand the "As-Is" IT support for ALC activities, we tried to understand the functional capabilities and limitations of its specific systems in terms of providing accessible, adequate information to support ALC management and decision processes. We also examined the interconnections and gaps between these systems, and analyzed the current architecture in which these systems are embedded.

Having examined the IT systems used by the various management processes related to ALC, we then sought to understand the extent to which the missing data and data accessibility problems noted above are related to limitations of these systems or to the "As-Is" IT architecture.

IT Support of Major ALC Activities

Our examination of ALC activity decisionmaking and management was included in our discussion of the various levels of decisionmaking (see those sections and relevant data-availability tables for specifics). Overall, the analysis showed that decision processes get limited support from the existing IT architecture, and that the level of support varies by activity.

When compared with the information needed, IT information support achieves varying levels of information availability and adequacy (see discussion of specific examples under the other levels described in this chapter). The main categories (along with an example for each) are as follows:

- **Accessible and adequate**: Graduation data and course length data in ATRRS.
- **Accessible but incomplete:** Unit schedule information in DTMS.
- **Accessible but questionable accuracy/currency:** Future force structure data.
- **Available but difficult to access:** Actual cost of producing and delivering the course (manpower and dollars).
- **Not available in any IT database**: Unit and NCO leadership strengths and key areas for improvement.
- **Available but not identified by activity managers**. There may well be data elements that could support ALC management but that are not used because management staffs are unaware of their existence.

The resulting "As-Is" IT architecture that supports ALC, as illustrated in Figure 3.9, is ad hoc, fragmented, and incomplete.

The figure shows services and systems (or databases) as circles with arrows between those that connect to each other. As can be seen, most connections between systems and between systems and ALC activities are "pairwise" (i.e., consisting of a single arrow that allows one system to communicate with one other system). Dashed arrows between circles in the figure represent systems that are minimally or unreliably connected, while circles that are not connected to others represent isolated systems. Dotted arrows represent information that must be validated or modified. Circles that may have information, but that activity staff are unaware or for some other reason do not use, are shown as circles with no arrow to the activity. Finally,

Figure 3.9
As-Is IT Architecture

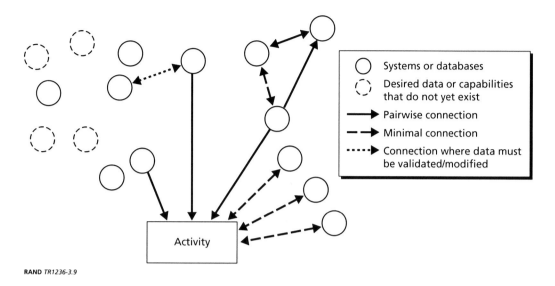

RAND *TR1236-3.9*

dashed circles represent desired data or computing capabilities that do not yet exist in any IT system. If anything, the complexity and irregularity of the figure actually understates the real-world problem that it depicts: Our research indicates that a major portion of the information needed by decisionmakers is not in any database or information system.

IT Accessibility

As we have seen, IT accessibility is a major issue. Staff personnel managing ALC must go to many IT systems to draw the information they need. Moreover, many of these systems are not easily accessed without special expertise and authorization. Additional expertise is often needed to interpret, transform, and transfer data to meet the user's needs.

A major cause of accessibility issues is the lack of effective and ubiquitous interoperability among IT systems. While ATRRS has been able to connect with other data systems, we found many IT systems that do not interface effectively with each other. For example, ATRRS, DTMS, and ALMS are all capable of maintaining information on student learning at some level, yet despite some efforts, their current abilities to exchange data are limited and unreliable or, in some cases, nonexistent. For a system to communicate with multiple other systems without undue human intervention, the program offices responsible for the two systems would have to forge a distinct pairwise connection between the systems, addressing a wide range of issues to ensure accuracy of data. Typically, different data systems will have been developed on different schedules for somewhat different, though potentially overlapping, purposes. Furthermore, not all systems adhere to the same policies regarding the safeguarding of Personally Identifiable Information (PII). Only ATRRS has universal usage, since it represents the "database of record" for Army institutional training. Efforts are continuing to link systems, but the difficulties are significant and progress limited. The lack of interoperability among existing IT systems reflects their stove-piped development, funding, use, and management.

Information Adequacy

There are many reasons why needed ALC information on IT architecture is not adequate. The most basic is that there is no effective collection of the information, such as ALC benefit data. These issues cannot be resolved by IT architecture improvements.

However, other issues are related to IT system design. The most obvious is when an IT system is designed without due consideration of the utility and the workload associated with its use. One ALC example is ASAT. It was designed assuming an ongoing process that did not exist, because the staff levels needed to implement the process were not resourced. Consequently, ASAT has the technical capacity to collect the data, but does not do so. An IT system is more than technology; it is a system that requires a human component. If a system is not used and checked, it will have absent or unreliable and invalid data.

Another example is cost information. For example, a key ALC need is an effective and funded POI and courseware development component. Funding needs to be provided not by the NCOA MDEP, but by a global training development MDEP. There is no real possibility that the workload and cost associated with POI and courseware development could be tracked in an IT database, because that would require the courseware development staffs and the NCO academy staffs supporting this effort to record the hours supporting ALC POI and courseware revision. Such a system could in theory be set up, but in practice it would not be unless directed by the school or center chain of command, and, even if it were, the validity of the information would be questionable without oversight. Moreover, in the absence of any quantification of POI and courseware currency or quality, it is not clear what conclusions could be drawn if the data were collected. We reiterate that the benefit of an IT architecture is a function of its human as well as technical components.

ALC Decision and Management IT Support Conclusions

Based on this examination of ALC Decision and Management IT Support we draw the following conclusions:

The "as-is" IT architecture provides limited support to ALC management and decision processes. Decisionmakers must collect information from a wide set of systems, and needed data are often not available from any system and either must be manually collected or assumptions must be made in the absence of the data. Sometimes, the IT system has the technical capacity to collect certain data, but the data are absent or incomplete. The reasons IT support of ALC is limited are varied and complex, so there is no easy or near-term promise for major improvement.

Nonexistent data and the lack of easy access to existing valid data among IT systems make it practically impossible to make fully informed decisions.

Even where information is available, it must be drawn from a wide range of IT services or databases, many of which are outside the training area. Some of the data would need to be analyzed to support ALC management and decision needs, and this could be a work-intensive effort, complicated by the fact that ALC IT systems continue to change. The most difficult issue is that many important information elements are not in any database or system. Decisions must be made based on assumptions and estimates. Given the rate at which operational requirements are changing, this presents major concerns.

Improved IT support of ALC management activities will take time and require effort. Because the causes of these limitations are varied and often outside the IT architectures, there is no reasonable potential for near-term, large improvement by means of IT architectural changes. Improvement of IT architecture should still be a goal; however, the expectations of benefit and effort required should be realistic.

Case Study Conclusions

In terms of the three elements of Institutional Adaptation, our examination of ALC management activities shows the following:

Support of Operational Readiness and ARFORGEN Processes

This case study's findings suggest that ALC needs a greater degree of adaptation to better support changing unit training readiness and longer-term leader development needs. ALC could also be changed to better align with ARFORGEN processes, specifically by increased use of DL and continued movement to execution at HSs. However, such levels of change would require significant alteration to course development and execution methods and processes, and a corresponding reallocation of resources. Such changes have proven systematically difficult to achieve.

The case study findings also lead to a conclusion that ATLD management processes and the lack of the information that supports them make it difficult to identify, prioritize, and support potential changes to ALC in such directions. A basic issue is the lack of the information necessary to understand fully the current training and leader development competencies needed by the soldiers who take ALC. Nor is there a system to gather the information needed to understand the areas where unit and NCO performance could be improved, or the nature of current unit training and leader development programs and their constraints. Such information is important to better shape ALC content and delivery methods for integration with ARFORGEN processes and thus to improve support of unit readiness and longer-term leader development goals.

Adopt an Enterprise Approach

Overall, we found that ATLD governance lacks structured assessment and integration processes for effectively adapting ALC in the context of an overall ATLD strategy. ALC decisions tend to be made from a narrow perspective of the NCOES program, rather than from one providing overall benefit to unit readiness and long-term leader development.

We also found that ALC governance is complex and involves many players, but that most decisions with regard to ALC courses are actually decentralized, and the ATLD governance structure has little visibility of ALC costs and student learning benefits beyond attendance and backlog data for assessment metrics. Instead, our examination indicates that decisions on ALC tend to *assume* the course provides certain benefits to the operational force. Moreover, the lack of information on unit training and leader development programs, needs, and constraints effectively precludes shaping and delivering ALC to align with unit needs.

An even more difficult issue is the integration of ALC learning objectives, acceptable course lengths and timing, and promotion policies. This issue is complicated by the somewhat different goals and shared ALC governance responsibilities of the manning and training communities.

Stewardship of Resources

There is no structured process to promote effective stewardship of ALC resources. Moreover, the underlying data on costs and benefits needed for effective stewardship is largely not available. The fact that resources to support ALC come from so many sources, managed by many different organizations with different priorities, complicates the achievement of an efficient ALC program.

Lack of true outcome measures especially complicates stewardship. The direct measure of "ALC graduates" is available, but not the course benefit to the graduate or the graduate's unit. Thus, the resource changes that affect these benefits are not visible in other than a general sense. Therefore, an important step to achieving more effective stewardship would be to focus more directly on the benefits—the outcomes—in terms of effects on NCOs and units, and to involve unit-owning commands in key decisions.

Based on the assessments discussed above, we conclude that achieving broader institutional adaptation goals could significantly improve the benefit of ALC to both near- and long-term unit readiness and to leader development, but that this will require significant systemic change.

Overall, our examination leads to the conclusion that ALC's management processes, while complex, are refined and accomplish many functions well. These positive capabilities should be maintained. However, it also shows that, while not easy, making changes to better align to the elements of Institutional Adaptation could improve the course's benefits and efficiency.

Implications

This case study involved one important ATLD activity among many. But many of the conclusions appear to apply more broadly. Specifically, the conclusions apply to the need for better integration of institutional and unit training to achieve a balance between near-term unit readiness needs and longer-range leader development requirements. The conclusions also apply to the need to have a more structured cost-benefit approach to managing ATLD programs.

In a larger sense, the case study supports a contention that better support of the greatly changing operational requirements within resource constraints requires major changes across ATLD programs. However, current management processes are not set up to make such changes.

CHAPTER FOUR

ATLD Program Support for Distributed Learning and Unit Training Programs

In the previous two chapters, a TRADOC institutional training course was examined and conclusions were presented about the potential benefits of applying an institutional adaptation approach to ATLD management processes. In this chapter, we review three recent ATLD-related RAND Arroyo Center studies and present conclusions about institutional adaptation of ATLD program management processes that arise from this review. These complement the ALC case study and broaden the basis for drawing conclusions about directions for ATLD management improvement. The first study focused on the Army's DL program, which is related to ALC and other institutional training and education courses. The second two studies focused on unit training and its support.

The DL study, completed in 2009, sought to help the Army assess the performance of The Army Distributed Learning Program (TADLP) to provide options for improving DL performance in both the near and long terms. The first unit-focused study, completed in 2007, had the objective of supporting Army efforts to develop and implement an effective training strategy for modernized BCTs by assessing the effectiveness of planned enhancements against changing training requirements and identifying the enablers that could best increase the effectiveness of these training strategies. It then drew wider conclusions about the effectiveness of the Army's system for making decisions about training enabler investments and directions for improving this system.[1]

The second unit training study, completed in 2008, focused on improving the Training Support System (TSS) management processes for providing products, services, and facilities to support Army training strategies.[2]

The Army Distributed Learning Program

Because DL issues were discussed at length in the ALC case study (see previous chapter), what appears below is a somewhat abbreviated summary of the research and focuses on the implications of this research to ATLD program management.

---

[1] This study effort is documented in Shanley et al., *Supporting Training Strategies for Brigade Combat Teams Using Future Combat System (FCS) Technologies*, Santa Monica, Calif.: RAND Corporation, MG-538-A, 2007.

[2] The Command General TRADOC and the Department of the Army's Director of Training requested this RAND study effort in 2006. This research was conducted by James Crowley and Michael Shanley.

The purpose of this study was to assess the performance of TADLP. The research documents the state of TADLP in 2007 and 2008 to establish a baseline against which future improvements of this program could be measured. In addition, the project outlined options that the Army could implement to improve performance in DL, focusing on strategic improvements that would allow the Army to leverage DL more effectively in the future.[3]

Established in 1998, TADLP is part of the Army's training and leader development system. DL seeks to enhance and extend traditional approaches to learning by making use of multiple means and technologies to enable the delivery of training and learning to soldiers and leaders wherever and whenever they need it.

DL capabilities, especially the ability to provide learning "anytime, anyplace," are becoming increasingly important in supporting the Army's training and leader development system. This is because, while training requirements are expanding, the Army's ability to increase (or even maintain) the time soldiers spend in formal training settings such as institutional schoolhouses is becoming more and more limited. For example, the flexibility of DL increases the potential of soldiers attending needed institutional training at HS during the Reset phase of the ARFORGEN cycle. Given this context, the Army has identified the need to transform training and leader development programs in a major way through increased use of DL.

Despite this growing recognition of the increased role to be played by DL, researchers found little movement toward its greater use. In fact, they found that TADLP had provided, at best, a modest ATLD benefit. Moreover, they found that resources for producing courseware within TADLP were both limited and declining. For example, in 2008, TADLP received only enough funding to develop a small fraction of the total institutional training requirement. Moreover, budget figures revealed that the production of DL courseware was receiving less, not more, emphasis over time relative to other ATLD programs. Finally, researchers found that the reasons for these outcomes could be traced, in large part, to areas where ATLD management needed improvement.

Approach

The study was done in three stages. In the first stage, data from FY 2006 (and informal spot checks in FY 2007–2008 to ensure continued validity) was used to assess the TADLP program against five measures of effectiveness for readiness-related courses: impact, efficiency, quality, cycle time, and responsiveness. In the second stage, options were developed for improving the courseware development program (as it existed through 2008) to address the areas of weakness identified in the first stage. Finally, in the third stage, options were developed and outlined for broadening the current TADLP to increase its impact, quality, and responsiveness, as well as for improving efficiency.

The study drew upon a variety of methods and sources, including reviews of relevant policy and program documents; analyses of Army institutional course management data and other databases; project-developed surveys concerning specific DL courses; interviews and focus groups with proponent schools, DL contractors, and TRADOC Headquarters staff; reviews of Army processes for developing courseware; and an analysis of the quality of selected IMI courseware.

---

[3]   This study effort is documented in Shanley et al., *Making Improvements to the Army Distributed Learning Program*, Santa Monica, Calif.: RAND Corporation, MG-1016-A, March 2012.

Conclusions

TADLP showed a need for improvement on all measures of effectiveness, and the research concluded that major changes in the way DL is managed are required both to improve TADLP and to better integrate it with other ATLD programs.[4]

Specific conclusions included:

TADLP courseware has had a narrow focus that limits its potential. The study found that the program had a relatively narrow focus on one approach to DL: using IMI—that is, stand-alone, computer-based instruction that does not involve any interaction between students and instructors.[5] The courseware focused on the learning and comprehension of facts, concepts, and procedures, often in preparation for resident training and higher levels of learning. This focus limited the learning objectives and the complexity of the skills that could be learned using DL. Blended learning approaches, which bring instructors into the loop and combine resident training with such DL methods as collaborative (synchronous and asynchronous) methods (and some online games and simulations as well), have been shown to apply to almost all classroom material, even complex subjects,[6] with little or no learning degradation. Collaborative DL also has the advantage of requiring many fewer training development resources to develop or change than standalone IMI. Examples of successful blended DL applications appear both inside and outside the Army.[7]

TADLP's focus is also limited to supporting structured TRADOC courses. The Army is implementing a broader learning capability called KM. A part of the KM initiative involves the delivery of training and learning products to support individual learning. To support this goal, a wide range of online collaborative forums and repositories has been established under the KM program to support individual learning beyond that provided in formal TRADOC courses. The issue is that TADLP and KM initiatives are not effectively integrated. While TRADOC devotes considerable resources to teaching institutional courses, developing TADLP courseware, and keeping courses current with new equipment, systems, and doctrinal concepts, there is no mechanism to keep the student current after graduation. Our examination supported the conclusion that better integration of TADLP and the KM program would benefit larger Army learning goals. For example, TADLP course modules

---

[4]   Subsequent to receiving RAND's TADLP report, the Army has made improvements with regards to some of the issues described in this section. However, TADLP has not received sufficient funding to enact major changes since the analysis was completed. Thus, the ATLD program management issues described below remain relevant.

[5]   IMI is a term applied to a group of predominantly interactive, electronically delivered training and training support products. IMI can link a combination of media, to include but not limited to programmed instruction, videotapes, slides, film, television, text, graphics, digital audio, animation, and up-to-full motion video, to enhance the learning process. IMI products include instructional software and software management tools used in support of instructional programs. IMI products are teaching and management tools and may be used in combination or individually. Used individually, not all IMI products can be considered interactive, multimedia, or instructional. However, IMI products, when used in combination with one another are interactive, multimedia, and instructional.

[6]   An example (described in the previous chapter) would be discussing student efforts to write OPORDs with the aid of student-to-student and student-to-instructor discussions online.

[7]   Use in asynchronous collaborative DL is the norm in civilian education, and in one Army example, the resident portion of the Special Forces Advanced NCO Course (ANCOC) was reduced from seven and a half to three weeks through the use of a combination of computer-based IMI and collaborative DL. In the collaborative portion of the DL phase, instructors and students interacted online asynchronously providing many of the benefits of classroom interaction.

could be made available of KM repositories and accessed without formally enrolling in the larger course.

Thus, the research showed that DL expansion beyond its current IMI and structured courseware focus could potentially provide large benefits to ATLD outcomes.

TADLP lacked a structured process for evaluation, assessment, and improvement. A major issue with TADLP was the lack of an overall process and supporting data for evaluation. None of the measures of effectiveness used in the assessment were in use when the study was started; rather, they had to be developed to support the project's research objectives. Usable data on the quality of the courseware largely did not exist, and many of the costs (e.g., for a school's contribution to courseware development) fell outside the TADLP program and were difficult to track. These shortcomings severely impeded the Army's ability to improve TADLP performance.

Designing effective improvement initiatives and verifying their success also lacked analytical support. For example, innovative new approaches to DL, such as blended learning, had no associated cost factors to support resource decisions connected with their implementation, and efforts to determine those factors had not been given strategic priority.

Thus, TADLP lacked basic metrics and an analytic process by which costs and benefits of the program could be measured, changes could be proposed, and progress could be traced.

TADLP lacked sufficient integration with other ATLD programs. Despite the fact that DL is a method of implementing institutional (and other types of) training, processes for creating DL courseware were not integrated with such training. As an example under TADLP, it took three years to develop a DL course, and three years to change it once it was developed. In contrast, residential courses typically are reviewed and changed at least yearly. The result was that many DL courses were not integrated with the resident courses they supported, or were out of date by the time they were completed.

In the last chapter, we saw that while NCOES did receive some emphasis within TADLP (i.e., more NCOES courses were chosen for funding), needed changes in DL to increase its effect and allow for transformational change were not addressed. Such changes would have required top-down guidance, resources, and orchestration, none of which were provided to support NCOES.

TADLP resource processes also lacked integration required to support major change. While TADLP is funded through its own MDEP, achieving a fully effective DL program would require supporting resources from a relatively large number of other sources, managed by many different organizations with different priorities. For example, in addition to working with the multiple MDEPs that fund institutional training, coordination with the training development MDEP would be required to ensure that training developers and Task Selection Boards were familiar with rapidly developing DL capabilities. Coordination with the Manning PEG would be required to build in instructor support for the various types of blended learning activities. MDEPs that have to do with simulations and gaming would need to be brought in on some occasions. Also, coordination with NETCOM potentially could be needed, as pipe bandwidth (needed for simulations and games and high-level IMI) varies by installation. Arrangements and resourcing also might be needed to allow soldiers to take the DL portions of the training from home (e.g., for facilities and computers). Despite this need for integration with other institutional training (and other) MDEPs, only informal and indirect mechanisms were present to coordinate and integrate efforts and funding.

TADLP Study Conclusions Relating to ATLD Management Processes

In terms of the three elements of Institutional Adaptation, our examination of TADLP shows the following:

- **Support of ARFORGEN Processes.** Expansion of TADLP has the potential to better support the Army's goal of having leaders conduct PME during ARFORGEN Reset, but the narrow focus of TADLP on IMI and the limited integration of DL with resident instruction has constrained movement in this direction. Likewise, better integration of TADLP, resident instruction, and KM learning delivery programs, along with more recognition of KM learning delivery as a priority TRADOC proponent function, could improve the Army's ability to keep operational-force soldiers and leaders current through all phases of the ARFORGEN cycle. Thus, our examination of TADLP indicates that improvement of ATLD program management process could provide for better support of ARFORGEN processes.
- **Adopt an Enterprise Approach.** TADLP governance processes do not provide sufficient data for a structured, metrics-based assessment of TADLP programs to support effective adaptation to current needs and constraints or to integrate them with the broader range of ATLD programs to provide the best possible Army-wide benefit.
- **Stewardship of Resources.** Again, as with ALC, much of the data needed to apply a structured cost-benefit approach to TADLP resourcing decisions either are unavailable or not easily available. A major factor is that the current structure of the Training and Manning PEGs do not provide for a full view of TADLP costs or benefits.

Brigade Combat Team Training Strategy Enablers

This study had the objective of helping the Army identify the most important enablers to support the Army's future training strategy for Maneuver BCTs equipped with Future Combat Systems (FCS) technologies. While the FCS program has been terminated and alternative programs for some of its specific systems—for example, tanks—are being reexamined, many of the key directions envisioned, such as increased intelligence, surveillance, and reconnaissance (ISR) capabilities, have been incorporated into current modular BCT organizations and concepts. So, while the study findings with regards to important enablers remain generally relevant, this summary focuses on findings and conclusions that are relevant to ATLD program management. The approach that was used to assess training enablers and identify the ones with the greatest potential benefit is directly related to ATLD management processes, and subsequent research shows that the findings and conclusions outlined in this summary are current and relevant.[8]

Approach Used in This Study

The study first examined BCT training programs in the baseline period (FY 2001–2002), prior to operations in Afghanistan and Iraq. This examination included the frequency, duration, and

---

[8]   Specific subsequent research includes the examination of TSS in the next section of this chapter and an examination of the Army's weapons training strategies conducted during 2012.

type of training exercises in these programs and proficiency levels achieved during collective training exercises at Maneuver CTCs.[9] Next, increases and changes were identified in post-2002 training requirements. Based on these examinations, gaps, or areas for improvement, between training requirements and likely unit training program achievements were identified. Next, an array of enhancements that had been identified for supporting BCT training strategies was assessed, and conclusions were drawn about their potential to improve unit training outcomes. Finally, potential ways to address the gaps were found, and recommended directions to improve Army processes for making decisions about training enabler prioritization were presented.

Baseline Training Programs, Capabilities, and Results

To establish a baseline, the content and output of eight heavy and five light 2001–2002 BCT training programs were first examined. This information was used to determine the training that BCTs were able to conduct, constraints on training programs, and areas in which these training strategies could be improved. An understanding of the nature of unit training programs in the baseline period remains relevant to current ATLD program management. With the Army expected to return to a more stable training environment than has been the case since 2003, an understanding of the level of unit training that was feasible and the activities commanders saw as important during this baseline period are valuable and relevant inputs to current ATLD management decisionmaking.

Baseline Training Heavy Battalion Programs. Table 4.1 shows data related to tank battalion training programs compared with the recommended number of events in the Army's Combined Arms Training Strategies (CATS).[10] The results for mechanized infantry battalions were essentially the same with a few exceptions, which are pointed out later in the section.

The table shows that heavy units performed far fewer live and leader training events over the period than recommended by CATS. Heavy battalions conducted only about three company- and platoon-level field training exercises per year, fewer than half of what is called for in CATS. They conducted a third fewer battalion-level field training exercises (FTX), including Maneuver CTC rotations. They also did very few fire coordination exercises (FCX) or command field exercises (CFX).[11] Units did conduct some field training events that are not in CATS, e.g., serving as opposing forces (OPFOR) for other units' situational training exercises and FTX.

---

[9] Maneuver CTC rotations are a major Army major training activity. BCTs in the U.S. deploy either to Fort Irwin, Calif., or Fort Polk, La., and maneuver as a brigade against a permanent Opposing Force (OPFOR). A large training support organization (Operations Group) supports this training and designs the training; role-plays higher, adjacent, and supporting organizations; tracks the training; provides feedback; and supports chain of command training assessments.

[10] CATS are doctrinal publications that provide commanders with a template for task-based, event-driven organizational training. CATS state the purpose, outcome, execution guidance, and resource requirements for training events. They can be adapted to the unit's requirements based on the commander's assessment. The CATS have been continually updated since the versions outlined in this section were reviewed. Also TRADOC and FORSCOM have developed and continue to refine ARFORGEN Event Menu Matrices (EMMs), which are templates of recommended unit training activities that supplement CATS. Reviews of more recent CATS and EMMs indicate that outlining more events than units were able to conduct in the baseline period remains an issue.

[11] Command post exercise and FCX are both types of leader training exercises. In a command post exercise, the unit's command posts are set up and skills in planning and executing an operation are exercised. An FCX is an exercise in which leader vehicles and unit command posts deploy to the field and conduct a limited-scale, live-fire exercise against a target array.

Table 4.1
Content of Tank Battalion Training Programs 2001–2002 Compared with CATS

Event	Annual Frequency: Actual Tank Bn Programs	Annual Frequencies: CATS
Gunnery Tables	2.6	2
CALFEX	0.4	1
Plt. Field Training Exercise	1.7	3
Co. Field Training Exercise	0.8	3
Bn/BCT Field Training Exercise	1.3	2
OPFOR Bn/Co	0.5	0
Co/Plt Virtual Sim Training Exercise	1.7	4
Leader Field Training Exercise	0.2	2
Bn/BCT Const Sim Supported Leader Training Exercise	2.3	4

NOTES: BCT = Brigade Combat Team, Bn = Battalion, CALFEX = Combined Arms Live Fire Exercise, Co = Company, Const = Constructive, Plt = Platoon, Sim = Simulation.

Tank gunnery was the only type of training for which tank battalions conducted more exercises than called for in CATS (averaging 2.6 compared with 2.0 per year). This difference occurred mainly because units scheduled make-up gunnery to maintain crew qualification rates in the face of crew turbulence.

Even though units performed considerably fewer field events than called for in CATS, on average they spent almost 100 days a year in the field because they spent many more days (and also drove more vehicle miles) per event than called for in CATS.[12]

Virtual and constructive simulations have a key role in the training strategies, but the number of actual simulations-supported events of this type that were conducted compared to CATS was even fewer than for field training.[13] The average tank battalion did about 60 percent of the CATS-recommended constructive simulations-supported exercises, and only 40 percent of the number of virtual training events. Mechanized infantry battalions did only about 20 percent of the number of virtual training events, the one exception to otherwise almost identical event averages.

Heavy tank and infantry battalions in 2001–2002 did a large portion of the HS tactical (non-gunnery) training during the period just preceding their CTC rotation. Thus they diverged from the doctrinal guidance, which calls for a "steady state" program across the training cycle.[14] Most of the tactical training occurred in this eight-month period (CTC rotation, including deployment, redeployment, and ramp-up) of a two-year cycle, with very little

---

[12] A major driver for determining the level of O&M funds allocated to units is the number of tank and infantry fighing vehicle miles driven.

[13] Constructive simulations represent systems and their employment through the use of extensive, complex mathematical and decision-based modules and statistical techniques. A constructive simulation is a computer program. The user inputs data to cause an event then gets the results. For example, a military user may input data on a military unit telling it to move and to engage an enemy target. The constructive simulation determines the speed of movement, the effect of the engagement with the enemy, and any battle damage that may occur. Results can be provided digitally or visually. Virtual simulations represent systems both physically and electronically. Examples include a video game or a cockpit mockup used to train pilots.

[14] See "Train to Sustain Proficiency" discussion in Chapter Two of HQDA, *Training for Full Spectrum Operations*, FM 7-0, December 2008.

occurring during the remaining 16 months. When considering skill decay and unit individual turnover, it was unlikely that units were able to maintain the same high readiness levels during the remaining three quarters of the cycle, when they trained less intensively.

Baseline Light Infantry Training Programs. The light infantry training programs reviewed differed substantially from those of tank and heavy infantry battalions. While heavy BCT programs peaked to a CTC rotation every two years, light infantry BCTs peaked to achieve Division Ready Brigade (DRB) readiness two to three times each year.[15] Light infantry training programs consisted of three distinct four- to eight-week cycles: (1) a support cycle with post-support, individual training, and similar activities; (2) a training cycle with collective training of squad through brigade activities; and (3) a deployment cycle in which the focus is on preparation for quick deployment and which includes practice deployment activities. In light infantry BCT programs, a CTC rotation was typically the major training event conducted during one training cycle. While some CTC preparation training took place, light infantry battalions did not conduct the extensive ramp-up CTC preparation programs we observed in the heavy battalions.[16] Also, while heavy brigades had limited simulation-supported training, light brigades did almost none.

Light infantry BCT training cycles were also shorter and trained a selected set of mission-essential task list (METL) tasks. Before a CTC rotation, heavy BCTs executed a progressive set of events, gunnery through BCT FTX, and trained on a more complete set of METL offensive and defensive tasks. Light unit training programs emphasized movement to contact against a guerrilla type threat and offensive military operations on urban terrain (MOUT). The 101st Airborne Division emphasized air assault operations. The 82nd Airborne focused on the airborne assault of a lightly defended airfield and its defense against light reaction forces. Thus, light infantry BCT HS programs focused on offensive tasks.

In terms of ARFORGEN readiness processes, the baseline heavy BCT programs, which focused on progressive readiness leading to a CTC rotation, are relevant to the Train-Ready portion of the ARFORGEN cycle, and the light infantry BCT programs are relevant to sustaining readiness in the Available portion of the cycle.

Baseline Training Program Output. As a part of this study, the output of the 2001–2002 training programs as seen in a study of CTC BCT Training Proficiency was also examined. To obtain the data to support this study, RAND collected questionnaire data from the CTC training cadre covering almost all of the organizations of a BCT at platoon level and above during the course of a rotation.[17] Each organization was rated on a range of key skills, tasks, and functions across the various operational functions after each mission. Up to 100 items were rated for each organization. Results were based on an average across almost three years of rotations. Key findings were:

---

[15] DRBs are BCTs that are expected to be at a high state of readiness and quickly deployable. Each division had one DRB always available.

[16] Heavy BCTs also had training cycles with three periods in each—a GREEN, a RED, and an AMBER. In a GREEN period, the BCT had priority for training areas and in RED periods they supported various installation requirements. In an AMBER period, the unit could train but would not have a priority for ranges, maneuver areas, or simulations facilities. The difference between the light and the heavy units was that during the GREEN periods (other than the one directly preceding the National Training Center rotation) heavy BCTs seldom trained higher than platoon level, while light infantry BCTs always included battalion- or BCT-level exercises.

[17] RAND also examined various data collected by the tactical analysis feedback facility for After Action Review purposes, including number died of wounds, weapons systems Operational Readiness rates, and Field Artillery and mortar firing logs.

- **For most units, proficiency on the large majority of tasks was inadequate the first time a task was performed.** This indicates that the HS training that these units conducted was insufficient by itself to achieve full training proficiency.
- **Overall, units appeared to achieve reasonable proficiency levels across many skills areas once the CTC rotation was completed.** This indicates that a CTC rotation is more than a test of unit training programs; it is an important component needed to obtain full training readiness.
- **The more frequently activities were conducted, the higher the percentage of units that reached proficiency.** This finding supports the belief that multiple iterations (i.e., a greater quantity of training events) are important to develop the ability to successfully perform combat skills under difficult conditions.
- **National Training Center performance results were less positive for maneuver battalions and BCTs than for maneuver platoons and companies.** In the higher-echelon units, fewer than half of the critical skills were ever performed at adequate levels by most units. This indicates a need for improvement in higher-echelon training programs.
- **Proficiency on synchronization and other key skills were lower for units at all echelons.** Finally, the data show that certain types of skills tend to be challenging at all levels, from platoon through BCT. These skills include direct fire skills, synchronization of combat multipliers, fire support execution, and intelligence exploitation. All of these are key to successful execution of future training concepts. Again, this indicates needed areas for improvement.

Changed Training Requirements and Constraints

Training requirements have changed since the baseline period in a way that places different, and in many ways more demanding, challenges on unit training programs. These include:

- **Persistent Conflict.** During the baseline period, operational deployment requirements were limited, but that is not expected to be true in the future.[18] The biggest effect is that unit time available for training, a key constraint during the baseline period, will be even more of a constraint for the foreseeable future.
- **Full-Spectrum Operations.** Maneuver units can no longer focus just on training for combat operations. They must be prepared to deploy to a wider range of operational missions and environments. This means that training and leader development activities must be more diverse, complex, and adaptable to changing theater requirements.
- **Modernization.** The centerpieces of modernization efforts are information technology—and specifically "internetted"—command, control, intelligence, surveillance, and reconnaissance technologies and precision fires. While these initiatives have enhanced operational capabilities, they also add to the requirements of the training system. In particular, they will affect training time. Fully leveraging these technologies requires training in complex analysis, planning, real-time decisionmaking, and rapid adaptation.
- **Modular Concepts.** Modular BCTs are designed to be more self-contained, having many formally divisional "slices," such as Military Police, included in their unit structure. Additionally, many command, control, and integration functions that were formerly performed

---

[18] This is true even though the Army no longer deploys units to Iraq, and operations in Afghanistan are expected to be far less extensive by 2014.

at divisional level have been shifted to the BCT level. Command and control training requirements will increase because the BCT takes on tasks that were previously performed at division level. Another consideration is that modular BCTs lack the branch-specific brigade, battalion, and separate company training oversight and expertise that was previously provided to engineer, military police, signal, and military intelligence units.

Conclusions from Review of BCT Training Programs and Emerging Requirements

The training system in the baseline period was highly successful, as shown by the initial tactical successes in Iraq and Afghanistan. However, that system required significant change to meet the changing operational demands in those theaters. Future full-spectrum training requirements will continue to change and generate challenges. The review above indicates that the degree of continued change needed remains, and continued training system changes will be needed. Moreover, it seems clear that the resources of time, manpower, and dollars will remain constrained and changes must be implemented within these constraints.

Assessment of Training System Enhancements

The next step was to identify key proposed training system enhancements. The Army's training system is complex and has a wide range of programs and activities that support its execution. In conjunction with the sponsor, a set of major training enhancements were identified and each enhancement was assessed against three key metrics:

- **Training quality.** The potential of the enhancement to increase the desired training effect, as determined by increased training event realism, complexity, and feedback.
- **Quantity of training events.** The potential of the enhancement to increase the number and duration of training events or the number of soldiers or leaders trained.
- **Adaptability of training events.** The potential of the enhancement to allow training events to be adapted to a wide range of missions, threats, conditions, and other considerations.

In addition, the benefits of each enhancement were evaluated in relation to "limiting factors," including constraints on unit time; technology/cost risk, that is, the risk that the technological advances would provide the desired training capability at an affordable cost; and the risk of less than full funding for the entire capability envisioned.

The assessments were based on reviews of requirements documents, discussions with training and materiel developers as well as Army staff responsible for training programs associated with the enablers, and the experience of RAND staff members working in related areas.

A summary of the training enablers examined and an assessment of their potential benefit are outlined in Table 4.2.

The results in this table are presented to show that a structured, systematic approach to making objective comparison of the relative benefits of a range of potential training enablers could effectively support more informed ATLD decisionmaking. A detailed assessment of the enablers in the table is available in Shanley et al., *Supporting Training Strategies for Brigade Combat Teams Using Future Combat System (FCS) Technologies*, Santa Monica, Calif.: RAND Corporation, MG-538-A, 2007. However to aid the interpretation of Table 4.2, a definition of each category appears below.

Table 4.2
Summary of Effect of Enhancements on Metrics for the Training of BCTs

Enhancement	Subsection	Likely Improvement in Training Capability Relative to Requirements in 2010–2016 Timeframe		
		Quality	Quantity	Adaptability
Home Station Live	• TESS	Some	—	Minimal
	• Targetry	Some	—	—
	• Ranges/Facilities	Some	Minimal	Some
	• Instrumentation	Some	—	Minimal
Maneuver CTC Modernization	• Instrumentation	Some	—	Minimal
	• Maneuver area	Some	—	—
Constructive Simulations	• Battle Command skills	Minimal	Minimal	Minimal
Virtual Simulations	• Ind/Operator/Maint Skills	Some	Much	Minimal
	• Crew/Squad skills	Some	Much	Minimal
	• Collective skills	Minimal	Minimal	Minimal
Laptop Simulations	• Leader skills	Some	Some	Some
LVC Integration and Tools	• LVC Integration	Some	Minimal	Minimal
	• Tools to support training	Minimal	Minimal	Minimal
Embedded Training	• Live	N/A	Minimal	N/A
	• Virtual Ind/Operator	N/A	Much	N/A
	• Virtual Crew/Sqd	N/A	Much	N/A
	• Virtual collective skills	N/A	—	N/A
	• Tactical Leader skills	N/A	Some	N/A
	• Constructive Battle Cmd	N/A	Minimal	N/A
	• IMI-based training	N/A	Much	N/A
Direct Training Support (HS)	• BCTC/CCTT	Some	Some	Some
	• ETC/BCBST	Some	Some	Some
Life Cycle Manning		Some	Much	Some
Institutional Training Init.	• Nonresident DL/SD/IMI	Minimal	Some	Minimal
	• AOT/JIT	—	Some	—
	• Reach back	Minimal	Some	Minimal
	• Battle Cmd Knowledge Sys	Some	Some	Some
MTPs/CATS/TSPs	• MTPs	Minimal	Minimal	Minimal
	• CATS	Minimal	—	—
	• TSPs	Minimal	Minimal	—
TRADOC Execution of FCS MBCT Initial Fielding		Some	Some	Some

NOTE: Ratings reflect usefulness of capabilities for the tactical training of modernized BCTs only (i.e., brigade-and-below training). They do not reflect an assessment of the value of these enablers for other training goals (e.g., for training above Brigade level or for CS and CSS support units).

- **Enhanced home station live training** are planned improvements to the Tactical Engagement Simulation Systems (TESS), ranges and facilities, targetry systems, and HS instrumentation systems.[19]
- **CTC enhancements** include a CTC modernization program, to include TESS, maneuver areas, and instrumentation systems.
- **Enhanced virtual simulations** are planned improvements in simulation-supported training, where real people are using simulated systems or equipment.

---

[19] TESS are systems that allow opposing forces in live training events to engage each other with direct and indirect fire systems and determine what the results would be if actual munitions had been used. Instrumentation systems provide for automated recording of force movements and engagement. This allows for a replay of the event to support After Action Reviews so that participants could see and understand what actually happened.

- **Enhanced constructive simulations** are planned improvements in simulation-supported training, involving the simulation of both the people/operators and the equipment they are using.
- **Simulation-based tactical skills trainers for leaders** are simulations (either virtual, constructive, or a blend) that can be delivered by means of a laptop computer and that have as their goal the training of tactical skills to individual Army leaders or small groups.
- **Integrated LVC** are planned initiatives to integrate different combinations of live, virtual, and constructive (LVC) simulations to improve training accessibility or quality, to plan LVC exercises, provide feedback, and increase the size of the training audience.
- **Embedded training** are efforts to embed training enablers in operational equipment.
- **Training manpower support for HS training** are planned increases in manpower resources that installations provide to support training events at HS.
- **Lifecycle manning** was an initiative under which a unit's personnel are stabilized for a period of 36 months.[20]
- **Institutional training initiatives** are proposed improvements in schoolhouse training to increase the availability of training and leader development information from the institutional domain.
- **Collective training support products** are proposed improvements in the primary products (current and planned) that TRADOC proponent schools provide to support unit planning and execution of collective training.

Conclusions

An ability to identify and defend the most important enablers will be key to the Army's success in effectively adapting its training system to meet future requirements. The Army must be able to field and sustain the set of enablers that provide the best possible overall training benefit within available training resources.

This assessment shows that making such decisions will be a challenge. All of these training enablers examined could provide potential benefit, but the amount varied greatly. At the same time, there was no "silver bullet" that would revolutionize training strategies. Of particular note is the degree to which many of the enhancements focused on technology with large potential but unproven benefits. In general, there was a tendency to overestimate what training technologies could accomplish.

However, the researchers found no structured process to assess benefits and costs across the range of potential enhancements to existing capabilities. They also found that enhancements tend to be considered independently, and often the ones selected have strong advocates. This leads to the conclusion that the processes the Army currently uses to select, fund, and prioritize training enhancements should be improved.

Improvement should involve evolving to a more comprehensive, analytic approach using metrics to support more informed decisionmaking. The metrics would include (1) metrics related to the effect of the enhancements on the quality, quantity, and adaptability of training; (2) cost metrics that allow a more complete identification of the full costs of given capabilities, and (3) field performance metrics that measure the effect of the enhancement and training on

---

[20] This initiative was never fully implemented, and the degree to which it might be reconsidered in the future is uncertain.

actual unit performance. These processes would also benefit from a better understanding of current training programs and constraints.

As a part of revised processes, the Army should consider using a spiral development process (continual observation, assessment, and analysis) to implement training enhancements. Many attractive enhancements may not produce the expected benefits or may have higher costs than envisioned. Therefore, early identification to reshape or cancel furthered implementation will be important to overall training system success and efficiency.

Supporting BCT Training Strategies Study Conclusions Relating to ATLD Management Processes

Compared to the three elements of Institutional Adaptation, our study of BCT training strategy support shows the following:

- **Support of ARFORGEN Processes.** Major changes in operational requirements since 2002 have driven the Army's move not only to the ARFORGEN cycle, but also to a major shift in the tasks, skills, and conditions that are the ends ATLD programs are to achieve to support ARFORGEN unit readiness goals. Moreover, changed technologies and organizational designs have also changed greatly since 2002. All of these factors affect and expand training requirements. BCT training programs are changing in major ways to meet these new requirements, and this generates major changes in the enablers required to provide the best possible support for these programs within available resources. However, this study shows that the Army has no process for gathering data on the nature and needs of unit training programs, nor a structured process for determining how to better support them.
- **Adopt an Enterprise Approach.** The Army lacks direct data on unit and leader performance effects on BCT readiness to support assessment metrics. Lacking such metrics, it is difficult for the training community to make structured prioritization decisions based on the overall unit readiness benefit of individual BCT training enablers. Instead, decisions of enablers are made individually and in isolation.
- **Stewardship of Resources.** The Army is not using a structured cost-benefit approach to select the best possible mix of BCT training enablers. Another issue this study points out is that the lack of realistic unit training strategies (type, duration, and frequency of training events) makes the process of deciding on the right type and amounts training enabler support haphazard. In a larger sense, the lack of cost and benefit data makes it difficult for the training community to make an objective case for its truly needed share of resources, or for the Army's leadership to understand risks associated with varying levels of ATLD program resource levels.

Improving the Army's Capability to Provide Training Support for Operational Forces

The second unit-focused study, conducted between 2007 and 2008, was a logical continuation of the first.[21] It involved a detailed examination of an important training program manage-

---

[21] While TSS supports all types of training in institutions and in units, this study focused on support of operational units.

ment process, TSS.[22] The goal of TSS is to provide a training environment that approximates the operational environment, providing commanders the capability to conduct tough, realistic, full-spectrum training. TSS fulfills the following functions:

- Develop, procure, and maintain training aids, devices, simulations, and simulators (TADSS), targetry, and instrumentation.[23]
- Build and maintain the maneuver areas, ranges, and facilities that house or store TADSS and targetry.
- Pay DA civilians and contractors who support training at installations.

As of 2012, TSS program budgets represented just over 1 percent of the Army's overall budget, and between 10 and 11 percent of the TTPEG.

TSS management involves the programming, budgeting, and execution of TSS resources as a part of the Army's Planning, Programming, Budgeting, and Execution System (PPBES).[24] It has three phases:

- **Execution.** Monitoring and adjusting congressionally approved TSS budgets in the current fiscal year (FY).
- **Budgeting.** Preparing the next FY TSS budgets for submission to Congress.
- **Programming.** Developing and refining TSS funding levels for the POM years.[25]

There are 11 specific TSS MDEP programs. These are organized into the five TSS program areas listed in the left column on Table 4.3.[26]

Note that the conduct of any type of training event requires resources from multiple TSS MDEPs and other Training PEG programs. In addition to TSS, other key training programs include:

- Operations and Maintenance, in particular OPTEMPO programs that provide the resources for fuel and repair parts
- TRADOC training support products that contribute to the commander's ability to plan and execute training.
- Service and training ammunition needed for force-on-force and live-fire training and which are provided based on the requirements outlined in HQDA, Pamphlet 350-38, *Standards in Training Commission (STRAC).*

---

[22] At the time this study was conducted, the TSS management process was not formally documented. The 2009 version of HQDA, *Army Training and Leader Development*, AR 350-1, has a chapter outlining the TSS. After review, we made some minor modifications in this section, but conclude that its basic content, findings, and conclusions with regard to overall ATLD management remain valid.

[23] TSS includes non-system TADSS. System TADSS, such as a flight simulator for a specific type of helicopter, are developed, procured, and fielded by the system Project Manager. However, once fielded, the sustainment of all TADSS, with the exception of embedded TADSS, is normally provided under a TSS program.

[24] See HQDA, *Planning, Programming, and Execution System*, AR 1-1, January 1994.

[25] The POM years are the six FY out from the budget year. In FY 2010 the POM years would be 2012–2017.

[26] The descriptions in this table were developed from a review of HQDA, *TSS POM 08-13 Requirements Briefs*, March 2006.

Table 4.3
TSS Training Capabilities

TSS Major Program Area	MDEP	Capability Provided
Battle Command Tng Support	TCSC	Battle Command Training Center (BCTC) operations
	TBWG	Constructive simulation development and procurement; BCTC facilities construction
	TCAT	Collective virtual simulation development and procurement
Soldier Tng Support	TBAS	Non system live TADSS and individual/crew virtual simulation development and procurement
	TAVI	Training Support Center (TSC) operations
Sustainable Range	VSRM	Range modernization (Major upgrades/new construction)
	VSCW	Training area and range operations
	TATM	Training area planning, management, and land repair and maintenance
CTC Modernization	TCNT	Non-System TADSS, instrumentation, and tng facilities for maneuver CTCs
Tng Support Infrastructure	WCLS	Sustainment of system and non-system TDASS
	TSAM	PEO STRI Management

- Manpower and facilities for Maneuver CTC training that are provided under the CTC program.
- Distributed learning courseware, technology, and learning management systems that contribute to the individual and leader skill learning that contributes to collective proficiency.

Approach

This study focused on improving TSS efforts to provide adequate training support to Army units. A research-based qualitative analysis approach was used, which included extensive dialogue with key members of the TSS community.

First, the current TSS was examined to identify its objectives, goals, and processes, and how the system functions. The next step was to assess the processes and identify areas needing improvement. Finally, specific directions the Army might take to improve the processes were identified and described.

The Training Support System Management Is Complex and Involves Many Key Players

TSS Community. Many organizations have key roles in the TSS management process. These include those listed below:

- **HQDA Deputy Chief Of Staff G3/5/7** exercises HQDA supervision responsibilities for defining training, education, and leader development concepts, strategies, policies, and programs.
  - The **HQDA DOT** assists the HQDA G3/5/7 in managing ATLD programs, including TSS.
  - The DOT's **Training Support (TRS) Division** develops TSS policy, provides the overall management of TSS, and plans, programs, and budgets TSS resources for DA G3/5/7.

- TRADOC **CAC Training** is responsible for the development and integration of the Army's training strategies and programs to train units.
  - **TRADOC Capability Managers (TCM) Live, Virtual, Constructive, and Gaming** align with the major TSS program areas and identify program requirements and support the planning, programming, budgeting, development, acquisition, and provision of TSS product, services, and facilities to the field.
  - **Army Training Support Center (ATSC)** is DA's Executive Agent for management of TSS. Additionally, TCM Live is under ATSC.
- **TRADOC Schools and Centers of Excellence** develop requirements documents for the TADSS needed to support their instructional courses and the training strategies of the units for which they are the proponent (for example, TADSS to support Field Artillery unit training by the Fires Center of Excellence at Fort Sill, Okla.
- **Program Execution Office for Simulations, Training, and Instrumentation (PEO STRI)** is an Army acquisition agency and develops, fields, and sustains TADSS, instrumentation, ranges, and targetry. PEO STRI takes approved requirements documents and moves these through fielding and sustainment.
- **U.S. Army Installation Management Command (IMCOM)** owns and executes installation support and operation of fielded TSS products, facilities, and services, except installations in Europe or Korea, or on Army National Guard installations.
- **U.S. Army Corps of Engineers (USACE)** builds and maintains installation TSS program facilities.
- **Army Commands (ACOMs), Army Service Support Commands (ASCCs), and Army National Guard Bureau (ARNGB)** (representing The Adjutant Generals [TAGs] of the states) are—along with TRADOC schools and centers—the TSS customers and participate in TSS management processes.

TSS Programming Process. The TSS programming process is shown in Figure 4.1.

**Figure 4.1
TSS Programming Process**

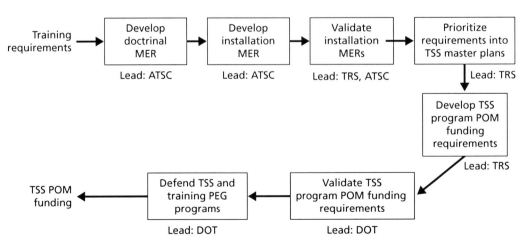

The programming process begins with development of doctrinal Mission Essential Requirements (MER) by the ATSC that document the TSS enablers needed to support specific types of unit training programs and institutional training courses.

At installation MER reviews, ATSC works with the installation's training staffs to develop and validate a list of training support enablers needed to support unit and school training at the installation. These become the Installation's MER.

The installation MERs are validated and shortfalls prioritized during a series of subsequent reviews. Army Commands (ACOM), Army Service Component Commands (ASCC), U.S. Army Reserve Command (USARC) and ARNGB, IMCOM, and proponent representatives participate in these reviews.

Validated installation MERs provide a basis for developing the TSS's Installation and CTC Master Plans, which chart a path to achieving full training support capacity.

Once given POM funding guidance, the DOT's TRS Division staff works in coordination with the broader training community to develop TSS POM funding plans that lay out dollar requirements by MDEP to best execute their master plans across the POM period. These funding plans go through a further set of management reviews, during which requirements are validated and from these critical requirements identified. At the end of the process, funded levels (affordable levels of critical requirements) are established for each TSS MDEP, and the generalized risks associated with funded levels are identified.

TSS Program Budget and Execution. During the budget period, based on updated guidance, the TSS MDEPs for the next year are refined into the budgets that are sent to Congress for review and approval. TSS program execution starts with the receipt of the fiscal year budget with congressionally approved funding levels. These budgets can differ from POM funding plans and submitted budgets, so additional staff efforts are required to support adjustment decisions.

During execution, funds are provided to the PEO STRI to develop, procure, and sustain TADSS. Products, facilities, and services are delivered to the commanders who execute training with the funding provided from the training support program budgets. The primary commands that execute these budgets are as follows:

- IMCOM provides the training support manpower that manages and assists in range and maneuver area operations, operates TADSS storage and distribution centers, operates simulations facilities, and coordinates the facility engineer assistance needed to maintain and perform minor construction projects to support training at both Active and Army Reserve installations. IMCOM does not provide this support in Europe or Korea, or at National Guard installations.
- The USACE designs and constructs ranges and other TSS facilities.
- ARNGB provides IMCOM-like training support funding to states. It also gets direct TSS funding to supplement that provided by IMCOM (at AC installations), PEO STRI, and USACE.

Constant change and adjustment are needed throughout the budget and execution years. Unforeseen requirements can arise, programs can be delayed or their costs can change, and priorities can change the amount of funding available. Ideally, the same type of coordination is needed for these adjustments as for the more deliberate POM development process. However, timelines are far shorter, and the degree of coordination possible is far less.

Assessment of the Training Support System Process
The assessment first defined four functions this process should fulfill given its goals and objectives, and then examined the degree to which these functions were fulfilled.

These functions are:

- Identify the TSS capabilities and resources needed to support unit training.
- Prioritize TSS needs and obtain and use TSS funding.
- Monitor and manage TSS execution.
- Adapt and respond to changes in requirements.

Identify the TSS Capabilities and Resources Needed to Support Unit Training.
The key component to this function is the development of doctrinal MERs, as outlined on Figure 4.1. These doctrinal MERS define the TSS required to support needed unit training activities in the POM years. A basic issue is that the Army has not developed a complete of set of training strategies outlining the type, frequency, and duration of the training events to be supported in the POM years to facilitate development of POM year training support requirements.[27] The CATS and weapons system training strategies examined did not document the full range of TSS support needed for training events; and the recommended set of events far exceeded what units have been able to do.[28]

A key issue is that the training community itself does not have access to the inputs it would need to develop training strategies for the POM years. Specifically, it does not have data on the nature of current unit training programs and their constraints, the levels of readiness provided by these programs and the areas in which improvement is needed, or the likely future operational requirements that could be expected during the POM years. In terms of TSS enablers, there is little data on levels of unit use and reasons for use or lack of use.

The complexity of reviewing installation MER worksheets and the demands of preparing for ongoing operations limits the ability to get unit chain-of-command customer inputs during installation MER reviews, especially from RC units.

Because training strategies do not define the full set of TSS capabilities needed to support unit training programs, and because and the data needed to understand the adequacy of current TSS support to unit training are not available, it was not possible to examine TSS programs and directly determine if the process was identifying the capabilities and resources needed to contribute to future training. Because of the training knowledge and experience of the participants, the capabilities identified may meet the requirements reasonably well. But better information certainly would make their decisions more visible and defensible, and a more structured management process would likely allow identification of a better balance of TSS and other training capabilities.

Prioritize Training Support Needs and Obtain and Use Funding. The factors that made it difficult to answer how well the TSS process identifies needed capabilities and resources

---

[27] The Army had two sets of strategies that define the needs for unit training programs: CATS define overall training strategies, and Standards in Training Commission (STRAC) guidelines define more specific weapons training strategies.

[28] As discussed previously, the CATS have been continually updated since the review outlined in this section was conducted, and ARFORGEN EMMs have been developed and refined to supplement CATS. However, more recent reviews of CATS and EMMs indicate remaining issues: outlining more events than units were able to conduct in the baseline period and that neither CATS nor EMMs documents the full set of TSS and other resources needed to support their listed training events.

also make it difficult to assess how well the prioritization function is performed. However, the review of the process and the MDEP briefs shows that the process does not provide the data needed for prioritization or to make a strong case for funding critical capabilities. A major issue is that the TSS MDEPs represent discrete training enablers, but most of the training activities they support require resources from a range of TSS and other MDEPs. Thus, it is not possible to directly link a reduced number of training events or the degree they will be less effective to a supporting MDEP. In an environment of severe cost constraints, inability to link training support capabilities directly to readiness can result in worthy programs being under-resourced.

Monitor and Adjust. During the budget and execution years, continuous TSS program adjustments are required, including internal shifting of resources between training support programs and adapting to changing requirements and guidance. Given the size and breadth of Army training programs and the range of products, services and facilities, this is a large and complex function. While decisions during budget and execution years would require about the same level of information and coordination with the TSS community as is needed for the POM process, they must normally be made much faster—sometimes in days or less. The number of organizations and the complexity of the TSS process mean that decisions are often made with less information and coordination than would be desirable. While the decisions made are the best possible under the circumstances, better and more information and any simplification of the process obviously would improve this function.

Adapt and Respond to Changed Operational Requirements. The TSS's ability to respond to unprogrammed operational requirements has been limited by the complexity of the system and the fact that it is geared to the more deliberate POM process. The ability to respond has also been constrained by material and system development and acquisition processes that are even more complex and deliberate. Mission-critical training support enablers to support unit preparations for operations in Iraq and Afghanistan were often provided much more slowly than desired by the operational community, and only with great effort. There have been successes, but these have been due to heroic efforts by all in the system to assist commanders, through creation of ad hoc procedures, and through the use of supplemental and unit operating funds to augment TSS budgets.

Assessment Conclusions

This assessment indicates that improvements are needed. This is not because the TSS programming process, as outlined in Figure 4.1 itself, needs improvement. This process is logical. While it is complex, complexity is not avoidable given the complexity of the PPBES and the large number of community members that must be involved in the process. Moreover, there is no evidence that the outputs of the process—TSS programs—are not reasonable, but neither is it possible to conclude that it is providing the best possible support to unit training programs within available resources.

This assessment also shows that the process lacks the full set of data needed for effective metrics and decisionmaking. This lack of available data did not allow for a definitive judgment on the how well the functions identified were performed. To the degree to which the TSS process is effective, it is the result of the experience and efforts of the participants. Overall, the assessment supports a conclusion that better data inputs would greatly improve the ability of the TSS community to manage TSS programs. Perhaps more important, better data inputs would increase the TSS community's ability to make the best possible case for its programs.

Directions the Army Could Take to Improve the TSS Process

Seven directions the Army could take to improve the TSS process were outlined. In many cases, the Army is already moving in these directions, and, in those cases, its actions are consistent with the findings and deserve priority.

Develop Overarching CATS to Facilitate TSS Programming. Development of overarching training strategies outlining the specifics of the training events to be supported in the POM years could improve the TSS in a number of important ways. Such models—if accepted by Army leadership as outlining a true baseline of the training events needed to achieve and sustain unit readiness—could serve as requirements documents for all types of needed training support capabilities and could become the integrated "Minimum Essential Events" list that the doctrinal MER would support.

Given the purpose of providing an overarching training resource model to allow the Army to budget for a "good enough" set of resources to support future unit training, training strategy models should have the following characteristics:

- **Define needed training events and key training enablers.** CATS should define the type, frequency, duration, OPTEMPO miles and days, and ammunition and key collective tasks trained of the events needed to reach the required levels of readiness.
- **Provide for full-spectrum readiness.** Strategies should be developed to provide full-spectrum readiness, understanding that adjustment would be needed for specific METL requirements for units assigned to prepare for a specific deployment mission.
- **Be overarching.** The strategies should be comprehensive and include all types of training events needed for full-spectrum readiness.
- **Be simplified.** The revised strategies could be built from the current efforts to develop the far-simpler ARFORGEN EMMs training templates and add only what is needed to provide the event detail and training enabler support requirements described above.
- **Be feasible.** They must set out requirements that units can reasonably be expected to accomplish. If the CATS outline more events than units can reasonably do, they lose their value as tools to program the right amount of resources.
- **Be adequate.** Even though POM strategies should be feasible, they still should describe strategies adequate to allow units to reach required training readiness levels. History suggests this is possible: the performance of baseline units at the CTC and in the initial phases of Operation Iraqi Freedom (OIF) and Operation Enduring Freedom (OEF) show that even though units did less than the recommended number of events, they were at more-than-acceptable levels of readiness for combat operational success.

Establish Systematic Mechanisms to Collect Unit Training Program Data. An understanding of the specifics of types, frequency, and duration of unit training events in actual unit training programs; why they were chosen; and the constraints that shaped their selection would enhance the training community's ability to develop and refine the training strategies described above that need to be supported and to identify important areas for training enabler improvement. However, there is no system in place for collection of these types of data.

Collecting data on actual unit training programs is not difficult, but will require in-place, real-time collection. This is possible if current capabilities and emerging initiatives and management systems are leveraged in a coordinated fashion.

Also key is an understanding of readiness trends across units, since these can indicate where systemic improvement is needed. As documented in the full report, Unit Status Reports (USRs) have widely recognized limits in documenting both needed improvements and training support resource limitations, and these limitations have made it difficult for the training community to make its case for resources. Initiatives were under way to improve training readiness reporting, and efforts to bring the training, personnel, and equipment status ratings into greater balance and make training ratings more objective are important and should receive a high emphasis.

- An enhanced system for operational assessments of unit training readiness could be coupled with the more objective ratings on the USR, both at the CTCs and during actual operations. Such assessments occur now, but the training community would greatly benefit from more structured efforts to identify systemic areas for improvement and then use this information to refine training strategies, describe needed training program changes, and specify the key training support resources needed to remediate training shortfalls.

Improve the Capability to Define Future Training Gaps and Critical Training Capabilities. While development of training support programs should be informed by current unit training programs, constraints, and needed improvements, these programs must also be developed to accommodate future operational needs. This means that the Army must define future operational requirements to the degree necessary to form the future training strategies to meet these requirements and to outline the training support enablers needed to allow those strategies to be effectively implemented.

Enhance Training Enabler Performance Measures and Metrics to More Directly Relate to Readiness. The measures used in TSS budgets grew out of the 2008–2013 POM, and the Army could improve them to provide a more accurate depiction of the gaps between resource levels and training requirements and to justify prioritization decisions. A central theme is the need to link training support to training readiness and the training events needed to achieve it.

The metrics for TSS MDEP performance generally fell into two categories—the percentage of planned events supported or executed and the percentage of scheduled, planned, or resourced projects completed. Neither of these describes the adequacy of support. The percentage of doctrinally recommended training events completed would be a more informative metric. Likewise, metrics could be developed to show the Army's overall ability to support strategies. For example, metrics could be developed to show the Army's capability to support CTC events in terms of quality, number, and ability to exercise the wide range of tasks and conditions needed to develop full-spectrum proficiency.

Refine and Institutionalize Processes to Respond to Operational Needs. The Army's responses to commanders' emergent training enabler requirements were at times slow and burdensome, despite major efforts by the entire Army to aid commanders preparing for deployment and in operational theaters. The problem was systemic. Assistance was generally provided through ad hoc systems and workarounds. The long-term need is to adapt processes to provide this support routinely, since it must be assumed that the Army will continue to have to react quickly to emerging operational needs.

Institutionalizing a systemic capability to respond to emergent needs is a complex issue, and without simple solutions. However, increasing flexibility is the key need. One example is by providing more training support manpower to help unit commanders plan, manage, and execute training events; another is by re-evaluating the constraints on use of Operations and Maintenance (OMA) funding for facility construction upgrades.

The Army should also study the processes that have been developed to provide responsive training support; understand what worked and what the shortfalls were; and, based on this study, develop and institutionalize a permanent capability.

TSS Study Conclusions Relating to ATLD Management Processes

Compared to the three elements of Institutional Adaption, this examination of the TSS amplifies the conclusions drawn from the BCT training and the other studies. It shows the following:

- **Support of ARFORGEN Processes.** Information to determine how well TSS programs are supporting ARFORGEN and operational force training requirements is unavailable, and many reasons contribute to this. First, there is little reasonably available data on the amount of TSS use. Even data on what specific unit training activities (that TSS supports) are being performed is unavailable. Moreover, for several reasons, it is difficult to get effective input from unit trainers on how TSS programs could be better shaped to support unit training programs. Unit training programs are very decentralized and until recently focused on meeting specific unit deployment requirements, meaning that there is likely wide variance. The consideration that commanders can use operational funds locally also complicates an Army-level understanding of what types of support are considered important by commanders and areas where they think the TSS could provide better support for the future. The complexity of TSS management processes and operational demands further limits effective participation of unit trainers in TSS decision processes. Lack of objective data on unit training readiness levels is another factor that prevents an informed judgment on how well TSS supports ARFORGEN processes.

- **Adopt an Enterprise Approach.** The TSS governance processes are complex and involve many players from a wide range of major Army commands. But as outlined above, they are without an effective understanding of unit-level programs and needs. The training strategies that the TSS supports outline a desirable—but likely unrealistic—level of training activities. These factors and a lack of objective training readiness data result in the training community having an incomplete set of readiness-related metrics. This, in turn, makes the process of making TSS decisions in the context of overall ATLD requirements highly dependent on the expert, but not fully informed, judgment of the training community. While these are likely reasonable, it makes TSS requirements hard to defend against the more objective metrics of the manning and equipping communities.

- **Stewardship of Resources.** For the reasons presented above, it is reasonable to conclude that the TSS community does not have the data to make informed decisions on the best possible use of available resources to support current and future unit training programs.

CHAPTER FIVE

Directions for Improving Training and Leader Development Management Processes

In this chapter, we first present our conclusions about the changed requirements and nature of the training and leader development strategies that ATLD programs support, and outline general areas where ATLD program management could be improved to better and more efficiently support these strategies. We then outline an overall concept, along with some specific directions the Army could take, to adapt its ATLD management processes to address these areas.

Conclusions

Based on our ALC case study, and other ATLD-related research, we draw the following conclusions:

ATLD Programs Have Changed, but the Need for Major Change Remains

ATLD strategies and programs have shifted. During the baseline period, they focused on preparing a generally uncommitted force for conventional combat. After the start of OIF, this shifted to preparing forces for counter-insurgency and stability operations in a training environment in which a major proportion of its forces were operationally committed.

The Army is now entering an era in which it must be prepared to face a far wider range of possible missions and mission conditions. Full-spectrum operations and the likelihood of more complex hybrid threats widen and complicate training and leader development activities, as do continuing organizational and equipment changes.

These changes must be made within the constraints of the resources available to ATLD programs. ATLD programs historically have been funded at less-than-required levels, and there is no reasonable expectation that this will change. In fact, given current Army budget outlooks, these resources are far more likely to decline. Moreover, time—both for unit commanders and in institutional training courses—will remain the dominant constraint, as the requirement to deploy forces, while greatly reduced, likely will remain significant for the foreseeable future.

Implementing Needed Changes Will Require Difficult Decisions

The changes needed are not a matter of going back to baseline strategies and programs. For example, full-spectrum scenarios at maneuver CTCs will be very different from the major conflict–focused scenarios of the baseline period, or from the counterinsurgency-focused scenarios in CTC mission rehearsal exercises for Iraq and Afghanistan. The ATLD resources needed to support these new scenarios will change considerably.

There are numerous examples of areas in baseline and current training and leader development strategies in which major change should be seriously considered. In the baseline period, about half of the field training time of heavy BCTs was devoted to tank and infantry fighting vehicle live-fire gunnery. While gunnery proficiency on these systems certainly remains a priority, given a changing threat, full-spectrum requirements, and improved weapons systems accuracy, re-examination of this allocation certainly seems warranted. Questions should be asked, such as, "What should be the balance between combat and stability tasks?" The answers likely will result in greatly changed training strategies and required resources.

On the institutional training side, our examination of ALC showed that potential changes such as increased integration of ALC with unit strategies and far greater use of DL and teaching of common leader skills at home stations in multi-MOS groups merit serious consideration.

Changes along these lines are not presented as an argument for any specific change. Rather, they are advanced to support the contention that major changes are likely needed, both in training and leader development strategies and the ATLD programs that support them. In a no-growth or declining fiscal environment, improvement in one area will require identification of other areas that can decrease in resources and many hard decisions will need to be made.

The Current ATLD Management Processes Are Not Set Up for Major Change or Flexibility

Current ATLD management processes were developed and worked well to sustain and make incremental improvements to successful, well-understood, and generally stable ATLD strategies. These complex processes were set up to validate and get resources for existing programs and to implement modest improvements. Because the larger ATLD system was seen as sound, limited emphasis was placed on establishing effective processes for examining overall system performance, or for analyzing options for making major changes from an overall system design perspective. As a result, the current ATLD Management System, while able to adapt in small increments, is not conducive to major change.

Future ATLD processes must also have increased near-term flexibility. The Army's efforts to adapt and meet emerging training requirements for operations in Iraq and Afghanistan were made possible because of major efforts, ad hoc processes, and use of supplemental and operations funds—and this area needs improvement.

Better Integration of Training and Leader Development Strategies and Programs Is Needed

No systemic processes are in place to integrate training and leader development strategies and programs for overall readiness benefit. Current processes focus on individual ATLD programs, with little look across all programs for overall benefit to readiness outcomes within available resources.

At the strategic level, the current practice effectively has separate training and leader development governance architectures, and this does not provide well for their integration.[1] Both a Training Strategy and a Leader Development Strategy have been developed.[2] While both have identified the desired aggressive ends (e.g. full-spectrum readiness, adaptable leaders), they are only beginning to come to grips with the difficult task of developing a consensus

---

[1]   The CSA's *Army Training and Leader Development Conference* provides a forum for facilitating senior leader dialogue on training and leader development issues, but does not have a decisionmaking or formal integration purpose.

[2]   This based on a review of HQDA, *Army Training Strategy*, April 2011; *A Leader Development Strategy for a 21st Century Army*, November 2009; and TRADOC Pamphlet 525-8-3, *Army Training Concept, 2012–2020*, January 2011.

concerning feasible ways and means for reaching these. For each, a wide range of initiatives has been outlined, but not how these fit together and, importantly, from where the time and resources for these initiatives will come.

Training and leader development generally are integrated activities. Leadership obviously affects how well units operate, and the training and motivation of their soldiers. Leader training and development occurs concurrently. For example, a leader going through a CTC rotation is being both trained and developed. In PME courses, leaders are being both developed and trained on technical and tactical skills.

Even within unit training strategies, there is a need for better integration. Gunnery strategies are developed by different organizations than the broader CATS, and in total, they outline a far more extensive set of activities than units can execute.

Integrated, well-defined ATLD strategies are very important inputs to effective ATLD program management. They outline what the individual programs are to achieve in the context of the overall readiness requirements, and provide a basis for reasonable allocation of resources across activities and programs.

The TTPEG's MDEP System Makes It Difficult to Make Decisions in the Context of Overall ATLD Benefit

The primary objectives of the ATLD programs are unit proficiency and leader competencies. These outcomes are achieved through direct training and leader development activities, such as ALC (and other PME) and CTC rotations. Thus, the logical management focus would be on direct ATLD activities.

However, the Army's process for managing resources, using MDEPs, defines programs at a much finer level of granularity and in a way that makes it difficult to manage major shifts in resources to support changing ATLD priorities. Some 122 MDEPs within the TTPEG support ATLD activity execution and are managed by the DA DOT. A few MDEPs in the TTPEG are "direct" in the sense that they focus directly on a key training or leader development activity. Many more are "support" MDEPs that provide resources (e.g., military manpower, installation support, ammunition, and training development) to a range of ATLD activities. Supporting MDEPs greatly outnumber direct MDEPs and often supply the majority of a direct ATLD activity's resources. For example, military manpower (instructor) support supplied by the manning community is far more closely related to effective ALC learning outcomes than are O&M resources provided by the core NCOES MDEP.

While support MDEPs provide resources to many activities, the resources allocated to direct activities generally are not directly visible. This makes it difficult to associate resources and costs at the activity level. Along with the large number of supporting MDEPs, this makes the process of coordinating, integrating, and justifying resourcing for direct ATLD activities complex and time-consuming.

In a period of relative stability, integration across training and leader development activities and supporting MDEPs evolves and becomes routine. But in a period of change, without precedents, no firm basis exists for determining what the adjustments should be.

Lack of Data Hampers Effective Stewardship of ATLD Resources and the ATLD Community's Ability to Make a Case for Needed Resources

The lack of activity cost data, discussed above, is coupled with the general lack of activity benefit data-objective measures of the effects of an activity's impact on unit training readiness or leader competencies. The result is that ALTD management processes lack a full capability to make objective decisions on the effective allocation of ATLD resources. Instead, decisions are made from the perspective of individual programs and types of resources, and not overall ATLD benefit.

Of even more importance, the Army has no system for objectively determining unit training readiness or leader competency levels, or ATLD areas needing improvement. Thus the ATLD community, compared to the manning and equipping communities, lacks the ability to quantify resource levels with readiness outcomes. The overall result is that the ATLD community lacks the data to make an objective case for needed resources, or to take a cost-benefit approach to allocation of resources to achieve the best possible ATLD outcomes.

Complexity and Lack of Integration Limit Operational Focus and Strategic Decisionmaking

The lack of a "big picture" view of ATLD program performance and needs, the focus on ATLD strategy components and individual programs, and the complexity of the strategic management process also make it difficult to focus on overall readiness goals. Many decisions are made in terms of component strategies, programs, and MDEPs, rather than in the context of what these mean to overall ATLD improvement.

These considerations also make it difficult to include effective operational force representation. Although FORSCOM and other unit-owning commands participate in many of these processes, process complexity and the number of advisory forums and councils make effective participation difficult. Moreover, these considerations mean that the potential for effective collaboration between the institutional Army staffs and operational force commands for time-constrained budget and execution year decisions is even more limited.

Areas for Improvement

Our overall conclusion is that current ATLD management processes, developed to manage incremental change, require fundamental changes themselves to support the greatly changed ATLD requirements. Based on our research, we identify three interrelated, general areas for ATLD program management improvement. These areas align with the Institutional Adaptation goals.

More Direct Understanding of and Focus on Operational Force Needs

Current ATLD processes focus on the resources needed to support training and leader development activities without a systematic process for looking at what these mean in direct terms to unit operational readiness. A major part of this issue is a lack of the information needed to understand where and how operational training readiness needs improvement, both for the near and long term. Another issue is that, for a variety of reasons, operational force commanders have limited visibility and influence on ATLD program decisions.

Increased Integration Across Strategies, ATLD Programs, and Other Program Evaluation Groups

Currently, there are separate training and leader development strategies, with limited mechanisms for integrating these to gain the best possible near- and long-term operational force readiness. Likewise, and partially as a result, integration across the specific ATLD programs and with the other PEG programs to provide the Army with a set of integrated capabilities to support operational force readiness is also an issue with current ATLD processes.

Development of a More Structured Cost-Benefit Approach to Making ATLD Program Decisions

Stewardship of resources would greatly benefit by taking a more informed and structured cost-benefit approach. Current ATLD program decisionmaking processes have little overarching structure and lack the cost-benefit information and metrics that would be needed to effectively put one in place.

Directions for Improvement

Our suggested directions amount to conceptual approaches to improve analytical support to ATLD program management and improved strategic governance architectures. The next two sections provide a detailed explanation of both directions.

Improve Analytical and Data-Collection Processes

In this section, we first describe what a more structured, systematic analytic approach might look like. In addition, we outline four specific management improvements to enable implementation of such an approach:

- Improve systems for collecting data on training and leader development programs' achievements, natures, and needs.
- Create improved mechanisms for managing by primary ATLD activity.
- Unify responsibility for data collection and analysis, and for supporting ATLD strategy and program management.
- Enhance ATLD and Army-wide IT architectures to improve data collection and analysis.

An Improved Analytical Approach

The approach we suggest is shown in Figure 5.1; it focuses on ATLD output improvement. Other approaches are possible, but any improved analytical approach would have similar basic elements.

The six steps in the improved process are:

1. Document ATLD activity outputs and costs.
2. Quantify leader and unit performance needs.
3. Identify and prioritize areas for unit and leader performance improvement.
4. Develop and analyze options for improvement.
5. Revise ATLD strategies.
6. Revise ATLD programs and activities.

Figure 5.1
Proposed ATLD Analysis Process

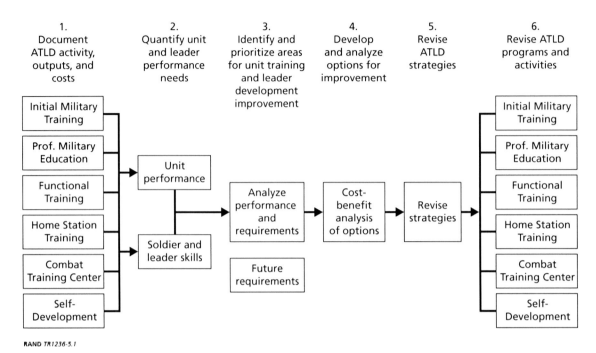

RAND *TR1236-5.1*

Step One: Document ATLD activity, outputs, and costs. The implementation of effective improvements in ATLD programs requires an accurate understanding of existing programs to establish a starting point (base case) from which change can be initiated. Current ATLD programs comprise a complex set of interrelated activities completed in units and institutions, with large variation across organizations. Currently, an overall documentation of these activities, which includes measurement of outputs and costs, is either lacking or not easily available.

Outputs can be measured in terms of the amount and characteristics of the training and education completed. A more robust capability to collect output data is needed. In the operational arena, no database records the number, type, and duration of unit home station training events actually conducted.[3] Even the training objectives of today's CTC training vary greatly by rotation because unit needs differ.

In the institutional arena, the number of graduates is well documented in ATRRS. POIs document the tasks trained, but POIs are often out of date. Outputs relate not only to how many students graduated and how many tasks were taught or trained, but also to how much learning occurred or what levels of proficiency were achieved. In many ways, the latter—activity outcomes—are more important than outputs, although both are valuable as decision support tools.

However, collecting data on outcomes such as achieved learning and training levels is far more difficult to accomplish systematically, and its measurement is often subjective. Thus, ATLD analysis often has to rely on output data as a proxy for full benefits. Potential activity

---

[3]   The Army has developed the DTMS to record such information. However, the use of this system by unit training managers has been limited and, at this stage, it does not serve as an effective database to achieve the purposes outlined in this section.

outcome data-collection mechanisms include improved end-of-course tests, surveys and structured interviews of graduates and participants, case studies, and structured training assessments of unit performance improvement during selected training events, such as during a CTC event.

Despite the challenges, an improved ability to collect and analyze data on learning and training outputs that would allow a better understanding of the contribution of training and leader development activities would support more objective ATLD program decisionmaking.

Also important is getting the views of the operational force customers on the relative benefits of various training and leader development activities and ways in which these can be improved. So is obtaining direct customer feedback on what training and leader development activities should have been completed but were not, and why. These data should become a major component of the analysis of reasons for shortfalls. Finally, customer input also could be used to help establish a priority for addressing shortfalls based on the criticality of the need.

Activity cost data are final elements in documenting ATLD activities as a base case. While cost tools such as ITRM and Training Resource Model (TRM) support the programming and budgeting processes, many important cost elements are not being collected, and in many cases data are not available. For example, while MDEP reports include DoD civilian and contractor costs, they do not include the cost of military manpower. Likewise, relatively little information is tracked beyond what is programmed about *actual* activity expenditures in unit training. For example, the Army can track the number of tank miles driven by a unit over a period of time, but not the training activities being supported by these miles.

Step Two: Quantify unit and leader performance needs. The first step involves understanding and collecting data on current individual ATLD program costs and benefits. The second step involves a data collection and analysis process to understand where ATLD programs need improvement.

Performance data needed to support analysis of possible ATLD program improvements would include specific individual and collective tasks and skills for which improvement is needed. Also, to identify Army-wide training and leader development issues, these data should be systematically gathered from a wide range of units in both training and operational venues. The data currently collected do not meet these criteria.

The measures used in the USR, the Army's primary mechanism for collecting unit training readiness data, are too broad to be used to identify specific improvements.[4] In the USR, commanders report their ratings of units' training proficiency on METL tasks and estimate the number of days needed to be fully trained. Commanders can also report critical resource constraints. However, these reports do not identify specific areas for improvement and thus cannot support the analysis process outlined above.[5]

---

[4]   See HQDA, AR220-1, *Field Organizations Unit Status Reporting*, April 2010.

[5]   There are also indications that USR data understate needs for ATLD improvement. In an analytical effort performed in 2002 we found that the training day assessments reported in the USRs (i.e., the number of days required to be fully trained in the unit's wartime tasks) were not affected by training and related activities that would be expected to change them. In particular, we found that USR ratings and training-day estimates are generally not affected by the length of time between training events, the amount of personnel turnover in the unit, the level of crew qualification, the amount of training events performed, or even the occurrence of a CTC training event. We also found that ratings on the training levels on METL tasks were higher in USRs than in briefings given at the CTC and to the chain of command. When we vetted these findings with knowledgeable members of the training community, we found a consensus that USR data were inadequate to support improvement of ATLD programs.

Likewise, the ATLD community gets little unit performance data that could support ATLD decisions from unit training activities at home station, CTC training events, or analyses of operational performance in Iraq and Afghanistan.

There is even less structure to the collection of data on areas in which leader competencies need improvement. The Army Research Institute (ARI) and the Center for Army Leadership (CAL) both conduct periodic surveys that provide some useful data in this regard. But again, these data are generally not specific enough to support a structured analysis process.

Other informal activities provide some insights into unit and leader performance and needs. For example, commanders and training staff at all levels spend much time visiting and talking with unit leaders. While certainly valuable, the results of these efforts are seldom systematically collected or used to support a structured analysis process.

Because ATLD improvement needs are typically not supported by structured analyses, major changes are often made without clear specification of the detailed aspects of the problems to be solved. For example, in our examination of ALC, we found no structured analysis of the NCO leadership shortfalls the transformed NCOES courses were to resolve.[6]

Step Three: Identify and prioritize areas for unit and leader development improvement. This step involves a structured analysis process that identifies and prioritizes critical areas for improvement as objectively as possible. The process and key inputs and outputs are outlined in Figure 5.2.

An improved understanding of current unit and leader performance gained from Step Two is an important input. However, even with reasonable improvement, these inputs, coming from multiple sources, will be fragmented, incomplete, and at their core be based on subjective judgment. The analysis process must be developed while accepting these limitations.

Figure 5.2
Step Three: Identify and Prioritize Areas for Unit and Leader Development Improvement

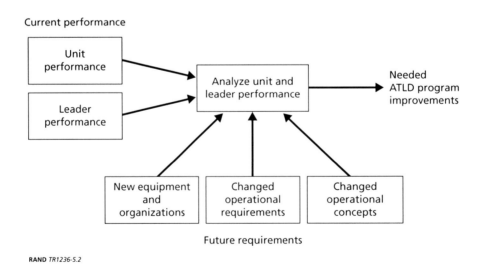

RAND *TR1236-5.2*

---

[6]   The NCOES transformation was generally based on the Army's 2006 *Review of Education, Training, and Assignments for Leaders (RETAL)*. In this review, the potential effects of the changed operational environment were examined, and an extensive set of interviews was performed to identify reasonable initiatives to strengthen NCO development. But this review did not include systematic, objective data collection or analysis to identify areas of NCO weaknesses.

Step Three also requires consideration of new or changed requirements. ATLD programs for POM years also must be shaped to produce units and leaders prepared for the wide range of possible future operational requirements. In this regard, effective collaboration with COCOMs and ASCCs to understand their changing operational needs and what these mean to prioritizing ATLD goals and objectives is important. The need for changed ATLD programs is also generated by new equipment, changes in organizations, new potential threats, or new operational requirements.

The importance and difficulty of this second part of Step Three are hard to overstate. Understanding and preparing for the next conflict, as opposed to the last, has always been a challenge. Given the wide range of possible operational requirements, the challenge is indeed significant, both in complexity and magnitude.

Thus, the analytical process in this step will heavily rely on qualitative analysis supported by some level of quantitative analysis.

Step Four: Develop and analyze options for improvement. The next step is to analyze options in order to determine how ATLD programs could change to achieve needed improvements in the most cost-effective way. This is complex, because unit training readiness and leader competency are the results of many activities, and effective improvement options will normally require a multi-activity approach.

For example, the Army has a multi-activity approach to improve capabilities against IEDs. Institutional, home station, DL, and CTC programs all have been enhanced to improve the capability to defeat IEDs, and material and doctrinal enhancements also are being examined as parts of an overall strategy for improvement.

The key point is that the ATLD community should develop a common understanding of critical areas for improvement, then develop a synchronized Army-wide plan for achieving it. Feasible options should be developed considering the full range of institutional and operational activities, and the role and potential benefit that each could reasonably contribute.

This is not to argue for a totally centralized approach. Decentralized initiatives are beneficial, and training program managers and executors should continue to assess and improve their individual programs. But they should also support or supplement Army-wide improvement initiatives.

In this process, options should be specific enough to allow for evaluation of benefits and costs. A structured cost-benefit analysis process should be used to evaluate and select the most effective option for improvement.

Because potential improvement options invariably will affect ongoing programs, negative effects must be understood and a conscious decision made to accept them as a price worth paying for the improvements to be gained. This means that improvement options also should include identification of resources and the effect of possible resource shifts on other programs. In developing options for improvement, considerations for maintaining current program effectiveness must be included. Our research indicates that such second-order effects are often not fully considered.

Enabling additional cost-benefit analysis on a basis that fits the (relatively short) DA decisionmaking cycle would benefit from investment in new analysis tools or "what if" capabilities that use the new data on outputs, benefits, and costs.

Improvement in the measurement of outputs and benefits (as described above) will increase the potential for making a meaningful assessment of benefits, costs, and effects. For example, our examination of ALC MDEP metrics showed that the implications of proposed funding reductions were not specific in terms of what these would mean in terms of fewer graduates or training events, less material covered, or lowered quality of the training.

Step Five: Revise ATLD strategies. A significant benefit of an improved data collection and analysis process will be that it could improve the ability of senior leaders to focus better on the larger picture and objectively make the difficult strategic level decisions needed to adapt the ATLD system to changing operational requirements. These decisions involve revising overall ATLD goals and objectives, not only in terms of what is of increased importance, but also what is less important. They also involve defining broadly, but specifically, how ATLD activities should be reshaped and how resources should be reallocated. With constant or declining resources, enhancing one area means taking resources from another.

As argued earlier in this report, adapting ATLD programs to support greatly changed operational requirements will require large ATLD system changes. Effective strategic planning will be key to effectively coordinating implementation of such changes.

While strategic decisionmaking will always have a major element of senior-level experience and judgment, the process we propose in the earlier steps will allow DA-level staffs to develop and present strategic course-of-action options with a more objective and complete outline of the advantages, disadvantages, and costs. It also could allow a better capability to monitor and adjust implementation.

Step Six: Revise ATLD programs and activities. This final step will require difficult decisions. Resources would need to be obtained and implementation plans developed.

The current ATLD program has been reasonably effective and has produced what is arguably the world's best-trained army. This means that changes should be made with careful consideration given to maintaining present quality. Given the complex interrelationship among activities, change in one program could have unintended consequences for overall system effectiveness. Therefore, major changes should be carefully considered, and made only when there are significant areas where performance must be improved or where operational requirements or resource availability have changed to the point where major adjustment is needed.

An advantage of a structured analysis process is that it can provide a defensible basis for justifying the resources needed to implement options in a total ATLD framework. The requirements that generated the need for change also could be revised into metrics to track and adjust implementation.

Implementation plans should include a process for assessment and revision. Plans for changing ATLD strategies and programs will almost always require adjustment. Sometimes, what seems like a feasible option for improvement will turn out not to provide the expected benefit. It also could have an unexpected impact on other programs or cost more than envisioned. Thus, the process for this step should be iterative, not linear.

Moving Toward Improved Analytical Support for ATLD Program Management
Our contention is that a more objective, structured approach to ATLD program management would benefit not only ATLD program decisions, but also senior leadership decisions on the reasonable amount of Army resources to allocate to ATLD programs. While such decisions always will be based on the military judgment of experienced, knowledgeable leaders, the approach we suggest would increase the Army leadership's ability to make informed decisions and support both POM planning and short-term ATLD decisions. Moreover, the approach

we suggest aligns to the institutional adaptation concept and to the Planning, Programming, Budgeting, and Execution System (PPBES) for managing resources.[7]

Fully implementing such an approach will require significant change. The systems to collect and analyze data to the degree we outline are not in place, and implementing them would take time. However, movement in this direction could start in the near term and have benefit, with an incremental implementation approach being used to continually improve data collection and analysis processes. Moreover, the Army can and should make this incremental movement using existing organizational resources, and not adding to them. The aim should be to make ATLD management processes more efficient and effective.

We next look at four specific directions the Army could start in the near term to begin implementing the more analytical data-based approach we are suggesting.

Improve Systems for Collecting Data on Training and Leader Development Programs' Achievements, Nature, and Needs During the ARFORGEN Cycle

Making the best possible decisions on training and leader development programs and policies requires a sophisticated understanding of the needs and nature of these programs. By "needs," we mean areas in which improvements are needed, which should be understood in the context of program strengths as well. By "nature," we mean what training and leader development activities are being done, their content, their rationale, and the constraints on these programs. Also important is getting the views of soldiers and leaders on the relative benefits of the various activities, and their ideas on ways that they can be improved.

It is possible to set up such a data-collection effort in a cost-effective way by taking advantage of ongoing efforts. Table 5.1 shows an outline of what information could be gathered in the context of the ARFORGEN cycle.

Data Collection During the Reset Phase. The goal of Reset is to provide for unit and personnel recovery and to accomplish the equipping, manning, and individual and leader training needed to allow collective training to begin within six months of return from deployment. Another goal is to accomplish required PME and functional training. There is an ongoing effort examining readiness programs for units in this portion of the ARFORGEN cycle, with the goal of improving Reset processes.

This effort could be enhanced to collect data on the full set of unit Reset activities and requirements, and on how much unit time and effort each requires. Indeed, some efforts of this type are currently under way, but our review of this program indicates additional information could be collected to support ATLD decisions.[8]

From the perspective of training readiness, a fundamental question is, when does collective training start? If the start is later than the goal of six months after return from deployment, other questions are: when did it start, what were the reasons for delay, and how could delays be reduced?

In terms of the ALC case study, key questions for surveys and focus groups would be: (1) How does sending leaders to NCOES affect unit Reset programs? (2) What is the length of time that can be reasonably allocated to this activity, balancing the benefits to the individual

---

[7]   See HQDA, *Planning, Programming, Budgeting, and Execution System*, AR1-1, January 1994.

[8]   We have reviewed documents outlining the Reset Pilot effort, discussed this effort with staff organizations at DA and FORSCOM, and attended a Reset Synchronization conference. Efforts to enhancing Reset processes continue, and the data-collection efforts we propose could improve their ultimate benefit to ATLD programs.

Table 5.1
ATLD Data Collection During an ARFORGEN Cycle

Pool	Data	Sources

Reset	• Unit activities—training and others
	• Training events conducted
	  – Type/duration/results
	  – Benefits and directions for improvement
	• PME requirements and completion—reasons for nonattendance
	• Manning and equipment levels achieved
		• Reset assessment efforts
		• Command Mission Training Briefs
		• Internal and chain of command assessments/After Action Reports
		• Training After Action Reports
		• ATRRS/TAPDB
		• Surveys and focus groups
		• USRs

Train-Ready	• Time between return and start of collective training
	• Reasons for any delays
	• Training events conducted
	  – Type/duration/results
	  – Benefits and directions for improvement
	  – Constraints
	• Compare against doctrinal training templates and find out why differences occurred
	• Training performance—e.g., crew squad quality, CTC performance
		• Mission Training Briefs
		• Internal and chain of command assessments and After Action Reports
		• CTC Take-Home Packages/Combat Trainer Questionnaires
		• Surveys and focus groups
		• USRs

Available	• Operational performance • Overall areas, and specific skills and tasks needing improvement • Directions training and leader development programs could be revised to support improvement	• CALL debriefs • Internal/chain of command assessments/After Action Reports • Surveys and focus groups

and the effect of his or her absence on the unit? Given the backlog issue, data could also be collected on the reasons NCOs requiring NCOES do not attend.

Data Collection During Train-Ready Phase. We found no systematic process for collecting data on the specifics of unit training activities during their dwell time. Yet an understanding of such activities (what is being done and why, the level of training proficiency achieved, and how the outcomes of training and leader development activities could be improved) would be of obvious value to training developers and to organizations trying to shape and resource their own programs.

Our previous research indicates that a program for collecting data on training activities conducted could be implemented with a relatively modest level of effort. A 2001–2002 RAND Arroyo Center effort examining potential readiness metrics was able to collect two years of detailed training activity data across 12 BCTs in five AC divisions.[9] This effort was made because we found that there were no data concerning the detail of home station training programs at DA or FORSCOM levels. The only way to get data was to go down to the brigade and battalion levels. The data were developed by collecting and analyzing existing training records, including Quarterly Training Briefs, training calendars, simulation center records, and unit briefings before CTC rotations. Also, brigade and battalion staff members were contacted and interviewed to verify and clarify the research results and to gain a general understanding of factors shaping their unit's programs. This research effort required about three months' effort

---

[9]   This effort and its finding and conclusions are documented in Shanley et al., *Supporting Training Strategies for Brigade Combat Teams Using Future Combat Systems (FCS) Technologies*, Santa Monica, Calif.: RAND Corporation, MG-538-A, 2007. In a related, unpublished research effort examining training strategies for National Guard BCTs, a similar effort was able to collect the same type of data for 15 brigades, thus indicating that such a data collection effort is also possible for the RCs.

by one researcher. Our conclusion is that such an effort is not only possible, but, if done in real time, could collect even more detail and yield even more useful insights.

Other activities could be added or mined to support the data collection effort. For example, feedback on "training and leader development areas for improvement" and "ways these could be improved" could be added as areas to be included in unit training briefings and After Action Reports. Installations could collect data on the usage of ranges, maneuver areas, individual facilities, and simulations facilities in a format that would give direct information on the type and duration of training activities by unit.

An important data element is "achieved training proficiency levels," especially in areas identified as needing improvement. While such data would be subjective, if collected in a systematic way it would still be of value to ATLD decisionmakers. A logical point for collection of such data is at externally conducted training events, such as at Maneuver CTC rotations and Mission Command Training Program (MCTP) exercises.

We realize there is justifiable concern over the collection of unit performance data from CTC events. The primary and obviously stated purpose of these events is training, and this primary goal must be protected. However, the purpose of the system we suggest would not be to compare units, but rather to evaluate the overall performance of ALDP programs and strategies. Thus, these data could and should be collected while protecting unit anonymity.

RAND Arroyo Center has successfully accomplished this in the past.[10] The main mechanism for data collection was a system where combat trainers at the Maneuver CTCs periodically completed short questionnaires assessing the proficiency of units on selected tasks and skills. The questionnaires were then transcribed into a protected database, from which analysis across a range of units could be conducted to examine performance systematically.[11]

Data on unit performance in Maneuver CTC and MCTP events and on potential suggestions for improvement are also captured in After Action Reviews and take-home packages (THPs) provided to units. These results could be systematically collected and analyzed to identify systemic training and leader development issues and areas for improvement. Currently, the CALL collects these Maneuver CTC and MCTP data and analyzes them to identify and prioritize issues that have Doctrine, Organization, Training, Leadership and Education, Material, Personnel, and Facility (DOTLMPF) implications.[12] However, the efforts of this organization currently focus largely on collecting and disseminating lessons learned from operations in Afghanistan, and not on more systematic analysis of DOTMLPF issues.

While CALL's priority emphasis on providing immediate, direct support to deploying and deployed units is unquestionably correct in the near term, enhancement of the system to collect and analyze training and leader development proficiency data from the CTCs and during actual operations would increase the Army's ability to make informed ATLD deci-

---

[10] See Hallmark and Crowley, *Company Performance at the National Training Center*, MR-846-A, 1997, and Shanley et al., *Supporting Training Strategies for Brigade Combat Teams Using Future Combat Systems (FCS) Technologies*, MG-538-A, 2007, for descriptions of how this system works, and examples of how the data derived can support training and leader development decisionmaking.

[11] Such a program would also be useful to have for home station training, especially as most combat support–type units do not undergo CTC-type training. However there are no feasible options for its collection. The CTCs' combat trainers provide the important capability for consistent evaluations, and no such capability currently exists for home station training.

[12] Such a process is proscribed in HQDA, *Army Lessons Learned Program (ALLP)*, AR 11-33, 2006. This regulation also has a requirement for the CTCs to report rotational insights and semi-annual trends to the CALL.

sions. Including questionnaire data along the lines described above would strengthen that data-collection system. Moreover, it would be beneficial to have such a system in place when the Maneuver CTCs and MCTP are able to fully focus their programs on training for broader full-spectrum training goals.

Data Collection During the Available Phase.[13] Data collection during actual deployments should focus on aspects of operational performance for which ATLD improvement is especially important. Obviously, such collection efforts would be of secondary importance to conducting operations, and would have to be structured to impose minimal or no real effect on operational units.

This could be accomplished by capitalizing on the capabilities of CALL team members embedded in deployed units. These teams develop observations, insights, and lessons (OIL); they could also identify and report on areas where training and leader development program improvement could benefit operational performance. Likewise, guidance on unit After Action Reports content could include identification of areas where training and leader development need improvement, and suggestions on ways improvement in these areas could be obtained. It is important that all these collection activities focus on issues that directly relate to operational outcomes.

Another good time for collection of data would be as soon as possible after redeployment. The thrust would be to have structured collection of leader feedback on areas in which their training and leader development programs needed improvement, and how these could be obtained. Indeed, TRADOC currently is conducting such efforts, but the focus is on organizational and equipment issues and tactical lessons learned. These efforts could be expanded to include greater emphasis on training and leader development.[14]

Finally, any system for collecting data on in-theater training and leader development strengths, issues, and suggestions for improvement would benefit from the direct input and review by the ASCC and COCOMs. In many ways, they are the ultimate customers.

Mechanisms for Data Collection. Much of the data needed to inform the Army's training and leader development already exist but not on any standardized, readily available database. The challenge is systematic collection. Moreover, many organizations—such as CALL, ATSC Liaison Teams, the Army's Asymmetric Warfare Group, BCKS, and efforts of individual TRADOC proponents—are already implementing related data-collection efforts. These efforts could be better integrated into a broader collection plan to provide additional data to support ATLD decisions.

Another component of the data-collection effort would be establishing a system of surveys and structured focus-group interviews to systematically collect the views of soldiers and leaders in operational units. Both the ARI and the CAL currently conduct broad survey efforts, and these could be modified to include items concerning training and leader development program effectiveness from across the Army.

Additionally, similar structured surveys and interview sessions could be executed at key points in the ARFORGEN cycle. Such points are early in Reset, immediately after Maneuver CTC and MCTP training events.

---

[13] If a unit does not deploy, the data collection methods for the Train-Ready phase could continue.

[14] Specifically, the Maneuver BCT TCMs at the Maneuver Center of Excellence orchestrate a program of visits to returning BCTs, which include surveys and structured interviews.

Again, there is great potential for developing a more enhanced and integrated ATLD data collection system. Such a system would certainly require effort and resources. But given current activities and efforts, a synchronized, standardized collection and analysis plan could be developed and executed with relatively modest additional resources. Such a program could be started relatively quickly and incrementally enhanced through an iterative spiral development process.

Unify Responsibility for Data Collection and Analysis and for Supporting ATLD Strategy and Program Management

The major theme of this chapter is that the ATLD program management and decisionmaking should be supported by a structured, systematic data collection and analysis process, as outlined in Figure 5.1.

Our research indicates that a centralized effort will be needed to implement and evolve this process, so a single, permanent staff organization should be assigned this function. This could mean forming a new organization or assigning the function to a current one, most likely with some augmentation. As discussed above, many parts of such an effort are currently ongoing, so this is not so much a matter of building a large new organization as it is of combining existing organizations into a more cohesive effort.

Here are some considerations in this regard:

Data Collection. Because data collection takes effort and resources, a collection plan must first be developed, with important information needed to support analysis and processes and responsibilities for collection identified. Collection plans would next be implemented and then modified as it becomes evident what areas need improvement and what data are possible to collect.

Assessment. The Army needs a more effective process for high-level assessment and analysis. It often seems that assessments are made from a narrow perspective (e.g., identifying an enhancement needed for ALC) as opposed to identifying specific areas of leader competency or unit training proficiency needing improvement, and then designing Army-wide ATLD program changes to achieve the improvement. While there is nothing wrong with individual program or activity assessment and decentralized improvement, at the Army level there should be a focus on identifying and mitigating Army-level problems, and this should include input from operation force commanders. As discussed earlier, Army-wide training and leader development improvement will seldom be achieved through a change in any single ATLD program—rather, it will take an Army-wide effort to develop an integrated, holistic, cross-ATLD program approach.

Option Development. Once ATLD issues are identified and prioritized, the next step is to develop feasible options for improvement. Again, effective options require a holistic concept involving a wide range of ATLD programs working in concert to improve an important ATLD output. Having a single responsible organization would be the best way to develop and shape such concepts.

Cost-Benefit Analysis. A key component of the processes recommended would be an effective cost-benefit analysis effort to enable more informed decisionmaking, and to determine trade-offs and effects both within the TTPEG and among the TTPEG and other PEGs. Often, we found that the analysis being done was within the narrow range of a single program or range of programs, and did not consider the full range of costs and benefits across ATLD and the Army.

Develop a Supporting IT System. The data collected and functions performed should be captured in an overarching ATLD IT system as quickly and completely as possible. Having a central organization responsible for this large and difficult effort is probably a prerequisite for meaningful success. We discuss IT architecture below.

Develop an Organizational Mechanism to Support Improved Data-Collection and Analysis Capabilities

Given the gaps in capabilities that appear to exist, and the complexity and difficulty of moving the current training and leader development system to one that better supports full-spectrum and persistent-conflict operational requirements, making the suggested modifications will be both important and difficult. As stated above, the Army needs a designated organization with significant internal capabilities to perform the data collection and analysis function. Even though this organization would support DA-level decision processes, the most feasible direction for improvement would be to assign a single TRADOC organization for the purpose of supporting DA-level efforts to integrate the wide range of training and leader development programs.[15]

The majority of the organizations supporting the functions described above currently exist within TRADOC. These include CAC-Training, CAC's CALL, CAC's Leader Development Center, and TRADOC's INCOPD. Each TRADOC proponent also has some capacities for supporting these functions in its branch and functional areas of responsibility.

TRADOC also owns a large portion of the Army's analytic capability under the Army Capabilities Integration Center (ARCIC) and the TRADOC Analysis Center (TRAC). ARCIC currently has responsibilities for DOTMLPF capabilities integration, and the new organization that we propose would support this ARCIC role by providing input on training and leader development improvement needs and how training and leader development could be used as part of the solutions for closing other Army capability gaps.

Performing ATLD data collection and analysis would not be so much a new mission for TRADOC as much as one of consolidating and enhancing existing capabilities. This means that the mission may well be able to be taken on within existing resources, or at least without significant increases. This is important, because it is not likely that TRADOC will get additional resources for such a mission.

While this organization would be under TRADOC and support its current training and leader development missions and roles, it would have a charter that effectively gives it a "direct support" relationship to the DA DOT, supporting its training and leader development policy and programming responsibilities.

A key role for the new organization would be to maintain close, continuous coordination with FORSCOM, USARC, ARNGB, and the six ASCCs that are assigned to COCOMs. Further, in conjunction with the DA DOT, the new organization would develop simplified, more visible processes for shaping and integrating various TTPEG programs for overall near- and long-term benefit to the operational force.

---

[15] The case could be made that this organization should be a separate Field Operating Activity (FOA) directly under DA G-3/5/7, similar to the Center for Army Analysis. But this likely would result in redundant and competing capabilities between this organization and TRADOC.

Enhance ATLD and Army-wide IT Architectures to Improve Data Collection and Analysis

An improved IT architecture has the potential of providing better support to ATLD analytical processes by increasing the amount of information available and by reducing the workload of collecting and analyzing that information. While some see it as having the potential for providing major long-term improvement, our conclusion is that pursuing such improvements, while necessary, will be of limited near-term value. In this section, we first outline issues that arise with regard to the support of ATLD program management under the current IT architecture. We then assess the potential for increased benefit, and finally present near- and long-term recommended directions for IT architecture improvement.

Issues with IT Architecture. While the current the IT architecture provides important support to management and execution of ALTD programs (see also our review of a number of those systems in Chapter Three), the overall level of support provided is limited. The reasons and causes for limited benefit are varied, and often outside the IT arena itself.

The fundamental reason is that much of the data and information needed for cost-benefit analyses are not in any Army-wide IT database or system. Examples with regards to "benefits" are provided in Chapter Three and in Table 5.1. Examples that focus on ALC and DL with regards to "costs" are also provided in Chapters Three and Four. In the collective arena, an example of cost data not in a database is the cost of supporting a brigade-level MCTP exercise. This activity involves a couple of hundred key unit personnel performing role-player and other support functions.[16] Moreover, this support is required not just during the exercise but during preparation for the exercise, which can last a week or longer.

A second major issue is that even where information is available, it can be difficult to collect. Data must be drawn from a wide range of IT services and databases, many of which are outside of the training area (e.g., databases dealing with Force Modernization). Since many of these systems are not easily accessed without special expertise and authorization, obtaining information can be difficult, and some of the data would need to be modified to support ATLD management and decision needs.

Furthermore, much information of potential use to ATLD management may not be located on a network, such as the spreadsheets used by TRAP participants. And even if it is on a network, the information may be in a format that defies easy exchange, collection, or interpretation. For example, units may post briefing slides of upcoming training activities on unit Web pages, but collecting such information from all Army units would create a workload beyond current staff capabilities.

Currency and completeness of data in IT systems is another major issue. Many of the systems are not being used, because the workload to input data into them is large and the staff to do this is limited and has other conflicting priorities. The same staffing constraints limit the use and oversight of IT systems, and IT systems that are not used and checked will have quality-assurance issues.

Another problem is a lack of interoperability of key data systems. For example, to support ATLD decisions and tradeoffs regarding unit training, it could be useful to improve the usage of DTMS and to improve interoperability of DTMS with the Training Ammunition Management Information System (TAMIS), as well as with the databases that store the associated

---

[16] An MCTP exercise is a Command Post Exercise performed by a TRADOC organization at Fort Leavenworth. For a further discussion, see Shanley et al., *Supporting Training Strategies for Brigade Combat Teams Using Future Combat Systems (FCS) Technologies*, MG-538-A, 2007.

MDEP information. These efforts could lead to a better understanding of the linkage between unit training events and needed training resources.

These and other issues involved in collecting data across training-related IT systems make it a major challenge to aggregate and analyze data to the degree needed to support informed ATLD program decisions.

Potential ATLD Benefits from Improved IT Architecture in the Near Term Are Limited. Our focus in this section is on improving IT architecture to support an enhanced analytical capability in Army-level ATLD decisionmaking. Our overall conclusion is that while current IT architecture provides some important support to ATLD management and execution, the support overall is limited. A complete, or even generally complete, set of data cannot be drawn from existing IT systems to support a "common training picture" and informed ATLD decisionmaking in the near term.[17] Because the causes for this limited support are varied and often fall outside the IT arena itself, there is no reasonable potential for changes to IT to lead to major improvements in management processes over the near term.

Despite the Challenges, IT Architecture Improvement Should Be Pursued. While IT architecture improvements may not, by themselves, lead to major improvements, they have the potential to simplify and streamline many ATLD governance and management activities by increasing the completeness, accuracy, and timeliness of the data available. Thus, they can yield worthwhile returns, which are likely to be increased as long-term IT improvements can be made. Effective process improvement will require additional efforts as well, including organizational, policy, or procedural changes. Given the fragmented and partial nature of data on current IT systems, the prospects of achieving the potential benefits of improved IT architectures to support ATLD program management would be greatly enhanced by placing this effort, as suggested earlier, under a single organization that is orchestrating the overall ATLD data collection and analysis process.

IT Improvement Approaches. Two strategies for IT improvement are possible and can be combined in many ways: Specific IT efforts aimed at providing or improving a narrow, targeted capability, and generic IT efforts aimed at providing or improving broad, widely applicable capabilities. Specific efforts (such as adding new functions to existing systems or creating new connections between existing pairs of systems) are likely to be successful if their target capabilities are reasonably easy to achieve and have a high payoff. Generic efforts (such as IT architectural redesign) are likely to be successful if the capabilities they provide improve the efficiency and/or effectiveness of a wide range of processes, and if they are achievable with reasonable investment and minimal disruption of existing IT capabilities. In the appendix, the two strategies for IT improvement are described in greater detail.

Plan and Prioritize IT Improvements. Systematic Army efforts to make improvements to ATLD data systems are still in their early stages (and further described below). Thus, we briefly summarize what would be required to plan and implement a program for IT improve-

---

[17] In HQDA, *Training Transformation Concept Plan*, 2008, the idea of a Single Army Training and Leader Development Enterprise (SADTLE) "Governance Dashboard" was illustrated in a figure showing a series of computer screens with a comprehensive set of data surrounding a map board in a command-post type setting. This dashboard was contained in several SATLDE briefings we received early in the project, along with the expressed belief that collecting and displaying a comprehensive set of data to support executive decisionmaking was a reasonable goal.

ments and to synchronize this effort with the data-collection and analysis processes outlined in the previous sections of this chapter.

Regardless of the approaches chosen for improving the IT architecture supporting ATLD management, a key element for success is specificity in selecting areas for improvement. Improvement across the full range of legacy systems may be neither possible nor needed. In this regard, IT improvement must be targeted to support the collection plan, discussed earlier in this chapter, which identifies the data and information needed to support the ATLD analytical process.

With regard to leveraging IT capabilities to support a collection plan and decisionmaking processes, an important step is in identifying and assessing potential IT resources. During this assessment, completeness, currency, validity, and accessibility of IT systems and databases would be examined and evaluated to determine the value potentially provided and the effort needed to utilize that potential. Identification and assessment will require a focused and likely an extensive effort. Our examination of the IT supporting ALC showed that just because a database or system exists does not mean that it is fully or even partially useful. Nor is it clear that all the potential IT resources have been identified.

Based on the assessment described above, collection plans and analysis processes should be revised to include IT systems and databases, as warranted. The capabilities and limitations of these systems should be considered, as well as the level of effort that will be needed to use and support each IT system or database.

The next step would be to improve capabilities to access and use these IT systems and databases. Our case study and related research both indicate that IT improvement carries risks and costs and can require significant lead time, often associated with a delayed return on investment. What might appear to be a relatively straightforward data collection effort can wind up requiring considerable effort, especially when integration is required. Therefore, careful consideration should be given to prioritizing efforts to improve IT systems and databases.

Create Improved Mechanisms for Managing by Direct ATLD Activity

Perhaps the most fundamental ATLD program management function has to do with defining, obtaining, and distributing the resources across programs to provide for the best possible benefit to unit readiness and leader competence.

In this section, we first describe issues that arise with regard to ATLD program management under the "As-Is" system. We then outline a direction to improve the visibility and support of direct ATLD activities.

"As-Is" Program Management Processes and Issues for ATLD. In the Army's "As-Is" program management process, MDEPs define Army capabilities at a fairly specific level. There are 122 MDEPs under the TTPEG, which resource the ATLD programs that support ATLD activity execution. In a general sense, MDEPs can be divided into two categories—direct (an activity that directly trains a unit or develops a leader such as CTC training or an ALC course) and support (to a range of direct ATLD activities).

As was shown in Chapter Three, resources from many other MDEPs, both inside and outside the TTPEG, are required for ALC execution. For example, as with most other ATLD activities, the most important resource for ALC, military manpower, is managed separately under the Manning PEG.

The fact that the supporting MDEPs themselves have constrained resources and multiple, independent priorities means that the managers of direct MDEPs are likely to have limited influence with regard to supporting MDEPs, especially when those MDEPs are in other PEGs. Moreover, the large number of MDEPs involved, and the limited visibility as to the level of resourcing needed from each one, makes it difficult to coordinate across MDEPs to support the direct activity.

Issues of seeing and coordinating the full resources needed for ATLD activity execution appear to apply generally. As an example, Figure 5.3 illustrates MDEPs supporting training at the Maneuver CTCs. Similar to Figure 3.7 in Chapter Three, Figure 5.3 shows MDEP and MDEP areas both within the TTPEG (on the bottom right) and in other PEGs (see boxes on the left) that support Maneuver CTCs. As with ALC, the resources needed to run the CTCs are funded out of a multitude of MDEPs (e.g., ammunition, OPTEMPO, and TSS) that are funded independently.

Outside the MDEP system, other important resources are needed to support CTC events, including unit time and unit support (for example, providing opposing forces and augmentee trainers). To benefit from a CTC event, not only must the unit spend time to prepare, but it must also move to the CTC and return to home station. All of these activities take unit time.

These examinations show that the "As-Is" process of allocating training resources across many ATLD activities is complex. One part of the complexity is the sheer number of MDEPs involved across the range of ATLD programs, which tends to give a fragmented view of the resources supporting direct ATLD activities. Another part of the problem is that MDEPs that produce intermediate products are too often treated independently from and not always completely integrated with the direct training and leader development activities they support. But the most important complexity is the large gap between required and available resources, and the process of closing them involves difficult choices. As a result, a large and time-consuming

Figure 5.3
MDEPs and MDEP Groups Supporting the CTC

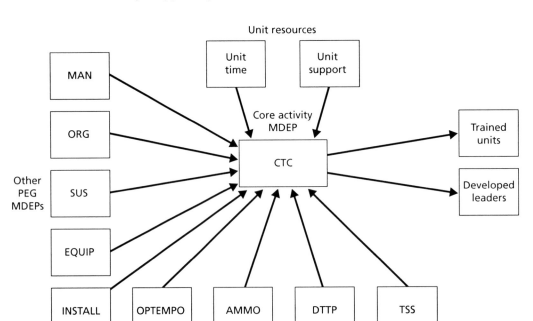

staff effort is needed to coordinate effective TTPEG integration and to defend individual MDEP resource levels.

Because the types and sources of the resources needed to support any activity are many and their contributions often not directly visible, the ability to define the effect of resourcing changes at the MDEP level is limited. This problem is compounded by the lack of an informed understanding of ATLD needs and ATLD activity benefits that are discussed above.

The current processes have evolved over time to provide reasonable support to a fairly well-defined overall ATLD system during the reasonably stable baseline periods. In our case study, we found the role of ALC to be well established, and that the program had reached a reasonable balance point in terms of value and resources required in the past. Only modest adjustments were required year-to-year, at least when resources and ATLD requirements remained reasonably stable. The same could be said of other training activities within ATLD. For example, home station and CTC training had, in effect, combined to form an overall unit training strategy. Basically, during baseline periods processes only needed to support year-to-year program sustainment and to make incremental adjustments to what was, by and large, a successful system.

But the demands for change on the ATLD management processes have been large and varied in recent years, making it difficult to continue to manage within the existing MDEP system. For example, the roles and relationships of home station and CTC training in the current ARFORGEN cycles are very different from those prior to the onset of operations in Iraq and Afghanistan. To support these changes, significant levels of different resources became critical to effective training. For example, a need arose for civilian role players, urban facilities, and IED simulations. Many of these resources had to be obtained outside the current ATLD program management processes. As another example, many unit commanders used unit operational funds to supplement those provided under the TSS MDEPs to quickly build the home station urban facilities and to hire the role players they thought were needed for effective predeployment training.

The recognition that further major ATLD program change will be required to reach larger full-spectrum ATLD goals suggests the need for a major reexamination of the roles, contributions, and nature of major training activities and how they are resourced. This, in turn, suggests that significant resourcing process changes seem needed to transition from an ATLD system that supported incremental change to one that can responsively support the major changes needed to support emerging training and leader development strategies.

Directions for Improving Mechanisms for Managing by Direct ATLD Activities. Our overall concept for an improved ATLD management process is one that focuses directly on outputs or, better, outcomes, like leader competencies and unit training proficiencies, and on the activities that directly result in these outcomes. Supporting this concept requires several things. The first is a structured process for determining where improvement is needed to support current operational requirements and to meet a fuller range of foreseeable likely, important full-spectrum operational contingencies. The second step is to understand how the direct activities contribute to achieving these outcomes. The next step is to determine how activities should change to achieve needed, affordable improvement.

To support this concept, DA would need to modify current mechanisms to enable management by direct ATLD activity. Management by direct activity would require modifying the current MDEP management processes to directly show the level of direct ATLD activity support from the range of supporting MDEPs.

Under this system, MDEPs that support direct training activities (e.g., DL or training development) would be considered under the major activity or activity they support. For example, DL courses would not be considered as a stand-alone capability, but as an important enabler of the critical institutional courses they support. Moreover, this provides greater visibility for supporting MDEPs, because it would allow their contribution to ATLD outcomes to be specifically cited.

While additional consideration would be needed, as described earlier, PME, IMT, Functional Training, Home Station training, Self-Development, and CTC training would be a good start for a listing of primary direct ATLD activities. Also, each of these primary direct activities should probably be divided further. For example, PME has Officer Education System (OES), NCOES, Warrant Officer Education System (WOES), and the Civilian Education System (CES), and the CTC activity has the Maneuver CTCs and MCTP. Other options are possible, but regardless of option there will be a need for integration across activities.

Our earlier Figures 3.7 and 5.3 both have shown how MDEPs might be aggregated by direct activity, using the Maneuver CTC and NCOES activities as examples. Supporting MDEPs would be affiliated with multiple direct ATLD activities.

Management by direct activity would include objective activity assessment. That involves determining how well the activity and its sub-elements are supporting the overall ATLD strategy, and the gaps and needs. It also would include developing options for improvement and performing cost-benefit analyses of each to develop a plan for improvement. This plan would be the basis for designing MDEP-level programs to support implementation and improvement.

Implementation of this approach would begin with the DA DOT assigning individual and organizational responsibility for each primary direct ATLD activity. For example, an MDEP at the core of the activity could be designated as a primary direct ATLD activity and be given commensurately higher supervision, responsibility, and level of staff support.

Under this new construct, organizational architectures, processes, and a set of forums would be established (or existing ones refined) to coordinate the development of an integrated set of program goals and supporting resources. As described above, many existing CoCs and working groups (e.g., Institutional Training, CTCs, and Home Station/Deployed training) can provide useful starting points for integration of activity programs and supporting resources. Moreover, it appears that the current DA DOT's staff organization could be readily modified to support an approach that manages by direct ATLD activity.

The current system for CTC Program Management seems to be an especially useful model for moving to such an approach, because, in most respects, it now operates that way. The CTC activity has a set of concrete activities on which it focuses. Its COC focuses on the full range of resources (e.g. TSS, military manpower, unit time, and augmentation) that support CTC events to level not currently true for other primary ATLD activities. It has organizational structures both at DA and CAC (the CTC Training Directorate) that work directly to provide analytical support and to develop a comprehensive "CTC Master Plan," which, in turn, directly supports POM development and budget execution.

Direct ATLD activities would also need to be supported with additional data systems and analytical capability, as described earlier in this chapter. For example, budgeting systems, such as ITRM and TRM, could be adjusted to make the specific costs of these direct activities more visible and easily available. Data also would be sought on the benefit side of the equation, with information collected on the scope of training outputs (e.g., number of training events conducted and duration of those events) and on the degree of learning achieved. The previous section described a possible approach for improved data collection. Further, analysis tools might be created that characterize the relationships among the component parts of the direct activity and that can assist with program integration.

Data on outputs, benefits, and costs by direct activity can, in turn, be usefully employed in the form of metrics that measure ATLD's success in meeting the goals of Army training strategies. For example, for home station training, metrics could be developed to show the Army's capability to support CATS in terms of quality, number, and ability to exercise the wide range of tasks and conditions needed to develop full-spectrum proficiency. Currently, metrics for ATLD MDEPs tend to focus more on program execution in relation to funded or planned training, rather than in relation to required training. A major issue for ATLD is objectively showing the effect of the gaps between required and funded levels. The inability to quantify readiness impacts makes it difficult for the TTPEG to compete with other PEGs. An example is the Equipping PEG, for which there is a much better connection between resource cuts and the consequences in terms of specific types of equipment that can be purchased or maintained.

Finally, management by direct program activity will require the additional data collection and analytical support, as described above, to achieve its objectives.

Evolve Emerging ATLD Governance Structures and Processes to Improve the Focus on Operational Force Readiness

Our examination of ALC and other ATLD programs at the strategic level led to the conclusion that further institutional change is needed. While the current decentralization of ATLD decisionmaking makes sense for purposes of implementing individual programs, such decentralization should be guided and overseen by a strategic management architecture that looks across programs and allocates roles and resources in ways that synchronize and best support overall ATLD benefit. This new architecture is needed because achieving a balance across a wide range of competing ALTD goals requires difficult decisions that need to be made from an Army enterprise perspective.

Below we outline additional directions that the Army should consider to enhance ATLD governance.

Re-establish an overarching Training and Leader Development GOSC to improve training and leader development integration. The current strategic management architectures should be modified to achieve better integration across ATLD programs. The current practice separates training and leader development considerations, with the TGOSC overseeing training strategies and the QLDR dealing with leader development. But establishing an overarching integration management process to support decisions across all TTPEGs, which support both training and leader development, still seems important to achieving better integration. Thus, a logical step would be to revise the charter of the TGOSC to one of providing "a management process to identify and resolve issues, determine priorities, and make decisions in support of Army training and leader development" and to "develop synchronized and integrated strategic recommendations for the CSA in support of Army Transformation and Force Readiness."[18] It would again become the Training and Leader Development GOSC

[18] This description of the role of the TLGOSC was in the August 2007 version of HQDA, AR 350-1, *Army Training and Leader Development*.

(TLGOSC), as was formally the case, and it could oversee and serve as an integrating mechanism for the full range of ATLD programs in the TTPEG.[19]

The TGOSC could be supported by a Unit Training GOSC that focuses on the collective and individual training events and activities that take place in units, and by the QLDR focusing on leader development activities in all three leader development domains (operational, institutional, and self-development).[20] Considering the important role of leader development in the operational domain and the importance of effective leader development as an enabler of unit training readiness, the QLDR should be co-chaired by FORSCOM, or at a minimum, FORSCOM should be given a high degree of influence and involvement.

Formalize FORSCOM's role in the Core Readiness Enterprise. To enhance the focus on supporting units in the ARFORGEN cycle, the Army should formalize FORSCOM's Readiness Core Enterprise role as the operational force's customer representative in HQDA, *Army Commands, Army Service Support Commands, and Direct Reporting Units*, AR10-87, September 2007.[21] A component of this function should include a stated role of representing the operational force at all forums involving training and leader development in units with a commensurate level of authority for influencing decisions and recommendations. Related and included in this formalization of FORSCOM's Readiness Core Enterprise role should be the authority, responsibility, and capability for developing a consensus on operational force positions on ATLD policies and priorities with other troop owning commands, including the ARNG and USAR.[22]

Streamline governance forums to increase the ability of operational forces to contribute to ATLD processes. The Army should consider modifying the CoC forums that underpin training and leader development governance and examine specific training and leader development areas.[23] We suggest shaping these CoCs along the lines of primary direct activities, as described earlier. This would reduce the number and streamline coordination processes, thus facilitating informed participation of unit-owning commands in the prioritization processes and focus on outputs. For example, having separate ammunition, TSS, and home station training forums developing issues causes focus to be on individual enablers of home station unit training rather on the total training strategies being supported. It also makes participation by unit-owning commands difficult. The starting question should be, what training events do units need for adequate training readiness, and how often and well must these events

---

[19] It should be noted that there is an Army Title 10 function of training, which includes leader development.

[20] The CSA directed that FORSCOM conduct a Collective Training Comprehensive Review to examine some key areas for transforming collective training to meet changed operational requirements and the current training environment. The ultimate intent is "of establishing the FORSCOM Commanding General as the leader of the Army's Collective Training Enterprise, much as the TRADOC Commander is the leader of the Army's Leader Development Enterprise." See CSA Memorandum, *Collective Training Comprehensive Review*, January 13, 2010.

[21] This regulation is dated September 2007 and "prescribes the missions, functions, and command and staff relationships with higher, collateral headquarters, theater-level support commands, and agencies in the Department of the Army (DA) for Army Commands (ACOMs), Army Service Component Commands (ASCCs), and Direct Reporting Units (DRUs)." As such, it is the main reference to the roles and responsibilities of the Army's commands.

[22] FORSCOM and other unit-owning commands participate and have a strong voice in the TGOSC now, but this elevation would reinforce and support the importance of understanding and supporting operational force needs in the management of TTPEG programs.

[23] The system of CoC and Work Groups is described in AR 350-1.

be done? This should then lead to an analysis effort to identify the enabler gaps and a process to fill the gaps, adjust the strategies, or clearly identify the risks connected with doing neither.

CHAPTER SIX

Conclusions and Implications

Overall Conclusions

Our overall conclusion is that ATLD management processes must change in a major way. Many might argue that while some improvements could be beneficial, the process overall is effective, and so there is no need for major change. However, an objective assessment argues for major change. There are major changes in the operational requirements that ATLD support and in the level of resources that will be available for ATLD programs. These changes generate a need for major changes in ATLD strategies and programs. We found that current ATLD management processes do not provide the structure needed to objectively make the major, difficult decisions needed.

In Chapter Five, we presented a framework for a more structured analytical approach supported by a greatly improved data collection and analysis capability. We outlined specific directions to support implementation of this approach. We also outlined directions that could be taken to evolve emerging governance structure and processes to improve the focus on operational force readiness.

While this approach and these directions will need further consideration and development, they are reasonable, feasible, and could provide significant improvement. We realize there are reasonable alternatives, but the overall point is that substantial changes in ATLD management processes are needed to focus more directly on operational force readiness, to integrate across programs for the best possible overall operational readiness benefit, and to foster improved resource stewardship.

Another main point is that a more structured, cost-benefit analytical approach could significantly benefit ATLD management and decisionmaking.

- A more analytical approach to support more informed leadership decisions would have benefit not only in terms of making ATLD program decisions, but in making decisions on the appropriate amount of Army resources to allocate to ATLD programs. While unit training levels and leader competency never will be measured to anywhere near the level of precision as unit manning levels or equipment fill rates, more objectivity would provide the Army's leadership a better understanding of the impacts of ATLD resourcing decisions. Once in place, it could support both POM planning and short-term ATLD decisions, such as reacting to changed budget allocations in the execution year.
- Adoption of this or a similar approach and movement in these directions could start in the near term and have benefits, with an incremental implementation approach being used to continually improve data collection and analysis processes. Moreover, the Army

can and should make this incremental movement using existing organizational resources, without adding to them.

Broader Implications

Our examination reinforces the obvious conclusion that achieving needed training and leader development levels involves decisions and actions both inside and outside the TTPEG and across all CEs. Manning, equipping, and installation policy and programming decisions affect training and leader development, and resources from all CEs and PEGs provide critical support to ATLD activities. Goals among PEGs can conflict and require difficult decisions about what is best overall, so cross-PEG and CE coordination is needed for overall ATLD effectiveness. An especially important area is synchronization between the TTPEG and Manning PEG.

A reasonable argument is that the current operating environment has given training and leader development increasing importance and thus improved the ATLD community's ability to support claims for resources. However, the ATLD community historically has had difficulties presenting objective analysis to support balanced resource decisions among training, manning, and equipping functions.[1] Absent a way to display the effects of decisions on readiness, analytically and objectively, too much risk in training is still likely to be accepted. The training and leader development community must be able to make its case in a way that provides better information on the risks and rewards of the hard decisions needed to take a synchronized Army Enterprise view across PEGs and CEs. This will, in turn, require that decisions across all PEGs and CEs be made considering training and leader development impacts, and that TTPEG decisions be made considering the broader readiness considerations.

---

[1] For example, see Defense Science Board Task Force, *Training for Future Conflicts,* June 2003; and Army Science Board, *Technical and Tactical Opportunities for Revolutionary Advances is Rapidly Deployable Joint Ground Forces in the 2015–2025 Era,* Summer 2000.

Strategies for Improving IT Support for ATLD Management

In this appendix we outline two strategies, one short term and the other long term, that the Army might take to improve IT support for ATLD management. The strategies are not mutually exclusive, and the Army could move to implement both concurrently.

Specific Short-Term Improvement in IT Architectures

One potential specific solution to problems of inaccessible data is to engage experts when necessary on a case-by-case basis. This is often accomplished by contacting a user or systems person associated with an unfamiliar system to access the system, then extract the desired information from it and interpret or transform that information as needed. This approach works, but it can impede decisionmaking, especially when repeated queries of an unfamiliar system are needed to explore data to look for patterns. A variation on this approach is for the users in an organization to hire a local expert who can access and interpret the data from a system that was previously unfamiliar to them. This can make sense if the organization has an ongoing need to use the system in question.

In some instances, improved data access and sharing would be easy to provide by using simple, inexpensive techniques. For example, as noted in Chapter Three, TRAP worksheets would be more accessible and consistent if they were Web-based instead of being shared as spreadsheet attachments in email. Quick, high-payoff IT solutions such as these are potential "low-hanging fruit," which can be identified, prioritized, and implemented without waiting for long-term solutions.

In other cases, specialized tools could be built to provide access to a stovepipe system from other such systems. This is warranted when the data held by multiple systems must be routinely combined, or when users of one system routinely require access to data from another system. In such cases, existing systems are typically made interoperable by means of individual connections between specific pairs of systems.

However, such pairwise connections can be problematic for several reasons. First, the creation of each connection requires considerable lead time, since it involves an agreement between the program offices responsible for two systems and the allocation of resources to craft a tailored interface between the systems. Second, the implementation of such tailored interfaces depends on the often-outdated technology that was used to build each of the two legacy systems involved. These interfaces must then be maintained and modified over time, as each system evolves along its own development path. For example, the ATRRS program office currently spends considerable effort maintaining its many connections to older legacy systems.

Because increasing numbers of old, unique interfaces must be maintained over time, this pairwise approach does not scale well as more and more systems need to connect to each other. Finally, this approach fails to provide the flexibility and agility needed to support dynamic decisionmaking, which often requires accessing new data or accessing specific data too quickly for a new pairwise interface to be responsive or too infrequently to justify building one.

Generic Long-Term Improvements in IT Architecture

A significant impediment to implementing an enterprise approach to training and leader development programs stems from the fact that existing systems have been developed as separate stovepipes. However, IT architecture throughout DoD is evolving toward a generic approach known as Service Oriented Architecture (SOA), which is designed to provide greatly improved interoperability among systems at reasonable costs.[1] The DoD's (and Army's) evolution toward SOA appears consistent with the Army Training Information System (ATIS) effort to support ATLD IT architecture. To understand the implications and potential advantages of SOA for ATLD, we therefore compared the existing "As-Is" IT architecture (Figure 3.9) with a "To-Be" architecture relying on SOA (Figure A.1). The "As-Is" architecture (on the left) was described in Chapter Three. Here, we focus on an explanation of the "To-Be" architecture.

SOA leverages Internet technology and World Wide Web protocols to realize its interoperability goals. It relies on a number of protocols and formalized specification languages, all of

Figure A.1
SOA IT Architectures

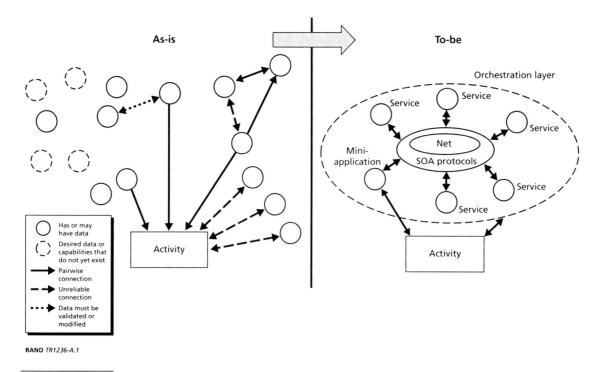

RAND TR1236-A.1

---

[1]   This trend is presented in DoD CIO, *Department of Defense Global Information Grid Architectural Vision: Vision for a Net-Centric, Service-Oriented DoD Enterprise*, Version 1.0, June 2007.

which are standardized and available to any SOA effort. The SOA protocols enable SOA services to invoke each other over the Internet, while the specification languages allow services to describe their capabilities and interfaces, specify their security and authentication policies, and publish the information that other services need in order to invoke them. The resulting specifications enable services to interoperate with each other, whether or not they were designed together or even with any knowledge of each other's existence.

SOA does not require tailored interfaces to be built between pairs of systems (as shown in the "As-Is" architecture on the left). Instead, it allows any system that has been restructured as an SOA service to dynamically discover any other service whose functional capabilities or data it requires and to connect to that service on the fly over the network, without the need for any prior agreement between program offices.

SOA provides the potential for flexible, scalable interoperability among IT systems, which should reduce the number of distinct systems that a training decisionmaker must learn to use and must access to piece together relevant data. Instead of learning to use a range of distinct systems, a decisionmaker should be able to create simple "orchestration" scripts that use the SOA environment to automatically identify and invoke appropriate services to perform a desired business process. In principle, SOA should enable Army training systems to interoperate with each other without prior agreement of any kind, thereby reducing the cost and lead time of creating specialized pairwise interfaces between them.

While there are obvious advantages of moving to an SOA, converting or replacing the legacy systems, services, and databases in the current ATLD IT architecture would involve significant cost. For SOA to work, IT systems must be converted into appropriate SOA services, conforming to SOA protocols and standards that describe each such service so that other services can discover and use them. Although SOA protocols and standards are already well developed and supported by DoD and the Army Chief Information Officer (CIO), it will take a significant effort to convert legacy systems to use the protocols and to design and provide appropriate service interfaces.

On balance, there are major advantages to transforming to an SOA approach. Moreover, DoD has invested considerable resources in developing an SOA environment for the Global Information Grid (GIG). We therefore suggest that the ATLD community should investigate the potential for moving incrementally to an SOA approach, and as warranted, begin moving in such a direction as soon as possible.

Indeed, several of the Army's future-looking efforts in the training IT area, notably the ongoing ATIS effort, already have tacitly adopted a To-Be architecture based on the new SOA paradigm. Although ATIS was still awaiting the establishment of a program office during our study, the ATIS working group and concept documents strongly suggest that ATIS is likely to pursue an SOA approach.

One view of ATIS is that if it is developed as a new, *ab initio* SOA system that replaces the functions of a wide range of existing IT systems within the ATLD environment, it will incur significant risk. Recent problems with implementing Enterprise Resource Planning (ERP) systems across DoD provide many examples (such as the Defense Integrated Military Human Resources System [DIMHRS]) of the difficulties of replacing multiple systems by a new "clean slate" system. A more likely (and in our opinion, appealing) scenario appears to be that ATIS will be developed as an umbrella system or architecture that provides an SOA environment and technical support for the incremental conversion of existing ATLD IT systems to SOA, to improve interoperability among them.

Fortunately, it is not necessary for all legacy ATLD IT systems to convert to SOA at once. An incremental approach, in which high-priority legacy systems (such as ATRRS and DTMS) are identified and targeted for evolution to SOA, would allow these key systems to interoperate with any new SOA services, whether they provide new capabilities or are conversions of existing legacy training systems. Similarly, ATLD should ensure that its new IT efforts (such as ATIS) conform to Army and DoD SOA standards and practices. ATLD should not move to SOA unilaterally, but should instead coordinate its incremental progress toward SOA with that of the wider Army Enterprise, since much of the information needed for ATLD decisions lies outside of ATLD systems themselves. For example, this would enable the ATLD community to rely on Army-wide SOA efforts to provide access across the full range of Army IT systems, such as those that maintain readiness, deployment, equipment, and facilities data. This seems feasible, since the Army Enterprise as a whole is moving toward SOA as well.

References

Army Enterprise Briefing, *October Army Enterprise Board (Read Ahead)*, October 2009.

Army Science Board, *Technical and Tactical Opportunities for Revolutionary Advances in Rapidly Deployable Joint Ground Forces in the 2015–2025 Era*, Summer 2000.

ATRRS Course Catalog, *BNCOC Common Core DL POI*, August 2008.

Center for Army Leadership, *2008 Leadership Assessment Survey, Final Report*, Technical Report 2009-1, May 2009.

Combined Arms Center Headquarters, *NCOES Transformation*, Operations Order 05-165A, July 2005.

Defense Science Board Task Force, *Training for Future Conflicts*, June 2003.

Department of Defense CIO, *Department of Defense Global Information Grid Architectural Vision: Vision for a Net-Centric, Service-Oriented DoD Enterprise*, Version 1.0, June 2007. As of September 13, 2012:
http://www.defenselink.mil/cio-nii/docs/GIGArchVision.pdf

Hallmark, Bryan W., and James C. Crowley, *Company Performance at the National Training Center: Battle Planning and Execution*, Santa Monica, Calif.: RAND Corporation, MR-846-A, June 30, 1997. As of September 13, 2012:
http://www.rand.org/pubs/monograph_reports/MR846.html

Headquarters, Department of the Army, *Planning, Programming, and Execution System*, AR 1-1, January 1994.

———, *Assignment of Functions and Responsibilities Within Headquarters, Department of the Army*, General Orders Number 3, July 2002.

———, *Army Lessons Learned Program (ALLP)*, AR 11-33, 2006.

———, *Review of Education, Training, and Assignments for Leaders (RETAL)*, 2006.

———, *TSS POM 08-13 Requirements Briefs*, March 2006.

———, *Field Organizations Unit Status Reporting*, AR 220-1, December 2006.

———, *Army Campaign Plan, Change 5*, April 2007.

———, *Army Commands, Army Service Support Commands, and Direct Reporting Units*, AR10-87, September 2007.

———, *Army Leader Development Program Charter Memorandum*, December 2007.

———, *Operations*, FM 3-0, 2008.

———, *Training Transformation Concept Plan*, 2008.

———, *Enlisted Promotions and Reductions*, AR 600-8-19, March 2008.

———, *Non-Commissioned Officer Education System (NCOES) Deferral Policy*, G-1 Memorandum, March 2008.

———, *Standards in Training Commission (STRAC)*, Pamphlet 350-38, October 2008.

———, *Training for Full Spectrum Operations*, FM 7-0, December 2008.

———, *Army Enterprise Board Charter*, May 2009.

———, *Army Training and Leader Development Guidance FY10–11*, Memo from CSA, July 2009.

———, *Management of Army Individual Training Requirements and Resources*, AR 350-10, September 2009.

———, *Institutional Adaptation*, briefing, November 2009.

———, *POM 12-17, Update Brief to the November 2009 Training General Officer Steering Committee.*

———, *Army Training and Leader Development*, AR 350-1, December 2009.

———, *Collective Training Comprehensive Review*, CSA Memorandum, January 13, 2010.

———, *Army Training Strategy*, April 2011.

———, *Unified Land Operations*, Army Capstone Manual, Army Doctrinal Publication (ADP) 3-0, October 2011.

HQDA—*see* Headquarters, Department of the Army.

Morey, John C. et al., *Best Practices for Using Mobile Training Teams to Deliver Noncommissioned Officer Education Courses*, ARI Report, No. A943005, 2009.

Roch, E., *SOA Costs and ROI*, 2007. As of September 13, 2012:
http://it.toolbox.com/blogs/the-soa-blog/soa-costs-and-roi-16794

Secretary of the Army/Chief of Staff of the Army, *Institutional Adaptation and Transformation*, Memorandum, January 2009.

Shanley, Michael G., James C. Crowley, Matthew W. Lewis, Ralph Masi, Kristin J. Leuschner, Susan G. Straus, and Jeffrey Angers, *Supporting Training Strategies for Brigade Combat Teams Using Future Combat Systems (FCS) Technologies*, Santa Monica, Calif.: RAND Corporation, MG-538-A, 2007. As of September 13, 2012:
http://www.rand.org/pubs/monographs/MG538.html

Shanley, Michael G., James C. Crowley, Matthew W. Lewis, Susan G. Straus, Kristin J. Leuschner, and John Coombs, *Making Improvements to the Army Distributed Learning Program*, Santa Monica, Calif.: RAND Corporation, MG-1016-A, March 2012. As of September 13, 2012:
http://www.rand.org/pubs/monographs/MG1016.html

U.S. Army Ordnance School, *91B30 Wheeled Vehicle Mechanic Advanced Leaders Course (ALC) Program of Instruction (POI)*, April 2009.

U.S. Army Training and Doctrine Command, *TRADOC Regulation 350-70, The Systems Approach to Training Management, Processes, and Products*, 1999.

———, *TRADOC Regulation 350-10, Institutional Leader Training and Education*, August 2002.

———, *A Leader Development Strategy for a 21st Century Army*, November 2009.

———, *TRADOC Regulation 10-5, Organizations and Functions, U.S. Army Training and Doctrine Command*, December 2009.

———, *TRADOC Pamphlet 525-8-3, Army Training Concept, 2012–2020*, January 2011.

U.S. Sergeants Major Academy, *Basic Non-Commissioned Officer Course Program of Instruction (POI)*, Common Core, June 2008.